GATWICK *IN PHOTOGRAPHS*

GATWICK IN PHOTOGRAPHS

CELEBRATING 90 YEARS OF LONDON'S SECOND AIRPORT

1936–2026

ADRIAN M. BALCH

The History Press

Front cover illustrations: Front, clockwise from top left: The original 1936 Beehive terminal building; A line-up of British Caledonian DC-10s in 1981; The present control tower opened in 1984; The South Terminal in May 1969 with British United Airways BAC One-Eleven 201 G-ASJE.

Back cover illustrations: Aerial view of the airport looking southwards in June 2021. (Ian Haskell); A trio of easyJet Airbus A320s queue up for departure against a stormy sky in April 2025; Your journey has come to an end – looking down runway 26. (Ian Haskell)

First published 2026

The History Press
97 St George's Place, Cheltenham,
Gloucestershire, GL50 3QB
www.thehistorypress.co.uk

British Library Cataloguing in Publication Data.
A catalogue record for this book is available from the British Library.

ISBN 978 1 83705 116 8

Typesetting and origination by The History Press
Printed in Turkey by IMAK

FSC www.fsc.org MIX Paper from responsible sources FSC® C111584

EU Authorised Representative: Easy Access System Europe
Mustamäe tee 50, 10621 Tallinn, Estonia
gpst.request@easproject.com

CONTENTS

INTRODUCTION

After having savoured the delights of plane spotting at London Heathrow Airport with my father in the early 1960s (all recorded in my book *Heathrow in Photographs*, also published by The History Press), I was keen to go further afield in my quest to see airliners elsewhere.

Gatwick Airport's traffic looked very interesting at that time, but alas I was too young to drive and it was more difficult to get to at that time. The M23 and M25 motorways were not built, so it would be a long cross-country road journey via Dorking, Reigate and Redhill to get there and, even then, rail fares were too expensive for this young schoolboy!

By sheer chance, I met a local aviation enthusiast who was older than me and drove a Mini Countryman, so early on a bright and sunny spring morning in April 1968, we set off for Gatwick.

On arrival, I was amazed at the variety of aircraft there on that day. There were smaller independent and charter airlines, not only British, but from all over the world, often using older airliner types that you would never see at Heathrow. Now fully established as 'the home of the independent airlines', several fledgling and start-up airlines moved into Gatwick with mostly second-hand equipment.

The same airliners that were seen in the colours of the world's major flag carriers at Heathrow during the 1950s and early '60s had now been passed on to secondary airlines to operate these veterans into Gatwick.

For example, Alitalia had passed on its DC-6Bs to SAM (Società Aerea Mediterranea), which were later replaced by some of Alitalia's former Caravelles. Some of KLM's DC-3s and DC-7Cs were bought by Martin's Air Charter, which became Martinair. Other second-hand DC-7Cs seen at Gatwick included some bought by Spantax from SABENA and others sold to Transair of Canada by Northwest Airlines. There were also start-up freight carriers such as Ace Freighters, who operated some pretty decrepit L-749A Constellations purchased from South African Airways and a couple of C-54 Skymasters.

The aircraft that could be seen ranged from light and executive aircraft in their own dedicated parking area through piston and turboprop airliners to British United's BAC One-Elevens and VC10 jets. Even military transport aircraft used Gatwick on personnel and freight flights shuttling between the RCAF bases in France and Germany. At Gatwick, the military personnel would then connect with charter flights by Canadian airlines.

The terminal building had three piers stretching out on to the apron and the tops of these provided perfect spectator facilities. In fact, although telephoto lenses were in their infancy, one was not needed to get good photographs of the aircraft parked alongside these piers – often they were even too close for the average photographer!

My friend and I made return visits over the following years to enjoy the delights of Gatwick Airport, and to record the sights and atmosphere of those glorious days, some of which I hope you can relate to in this nostalgic timeline of photographs and memorabilia, whatever age you are.

It is clearly not possible to mention all the airlines that have come and gone at Gatwick over the years, or to illustrate them, as this book is just a pictorial timeline snapshot of the scenes there over the years, so apologies if your favourite airline is not mentioned or depicted.

Today, Gatwick is the UK's second largest and one of Europe's top ten airports, with 43 million passengers annually. As an important area of national infrastructure, it generates £5.5 billion a year, supporting more than 76,000 jobs supporting both the economies in the nation and region. Nearly sixty airlines fly from the airport to more than 150 short-haul destinations and to more than fifty long-haul locations. There are some fifty-five movements an hour, landing and taking off from London Gatwick's single runway, claimed to be the most efficient in the world.

Located 28 miles south of the UK capital, Gatwick is extremely well-connected, facilitating more than a quarter of England's population (15 million people) – including all of London – who need to travel less than one hour away by road or rail. An ambitious plan to become a net zero airport by 2030 involves a six-year, £2 billion sustainable growth programme, increasing capacity by incorporating the airports' Northern Runway into regular use, alongside its current main runway. Vinci Airports manages 50.01 per cent of the airport, while Global Infrastructure Partners manage the remaining 49.9 per cent.

ACKNOWLEDGEMENTS

This book would not have been possible without the help of some good friends who supplied photographs and memorabilia, to whom I express my sincere thanks, in particular Richard Andrews, Mike Axe, Brian Burrage, Rebecca Byng, London Gatwick Media Relations Manager Colin Cooke, Martin Fenner, Ian Haskell, John Hughes, Tom Singfield, Richard Vandervord and Stephen Wolf.

Any photographs uncredited were taken by the author or from the author's collection. Where the original photographer was unknown, please accept my apologies. Every effort has been made to check necessary permissions, but in the event of any omission, please contact the author, care of the publisher.

ONE

1930S–40S: IN THE BEGINNING

In the beginning Gatwick was little more than a recreational aerodrome between Gatwick Manor and Hunts Green farmsteads, alongside the racecourse that was there at the time. The airfield started life in 1930 as the Surrey Aero Club, a small enthusiast's flying club, but it did not stay this way for long. As an airport, Gatwick began operations in the mid-1930s (at a time when Heathrow was still countryside). It was licensed as a public aerodrome in 1934, to enable a regular air services to Paris to commence, also acting as a relief aerodrome for London Airport at Croydon. First scheduled flights were by Hillman's Airways to Belfast and Paris with DH84 Dragons. The legacy of Hillman Airways can be traced to British Airways, which operate from the airport today.

In 1934, construction began on the Beehive, the world's first circular terminal building. The following year, Gatwick's train station was opened to serve the new airport.

↑ Gatwick Airfield was established in 1932 with the Redwing School of Flying. Robinson Redwing 2 G-ABMJ can be seen on the right.

↑ By 1936, the airfield had expanded to include the hangar owned by Air Travel Ltd, with Fairchild FW.24 Argus G-AECO in the centre.

The decision is made to expand the airport in 1936 and construction begins on the Beehive terminal building with control facilities.

1930s adverting poster.

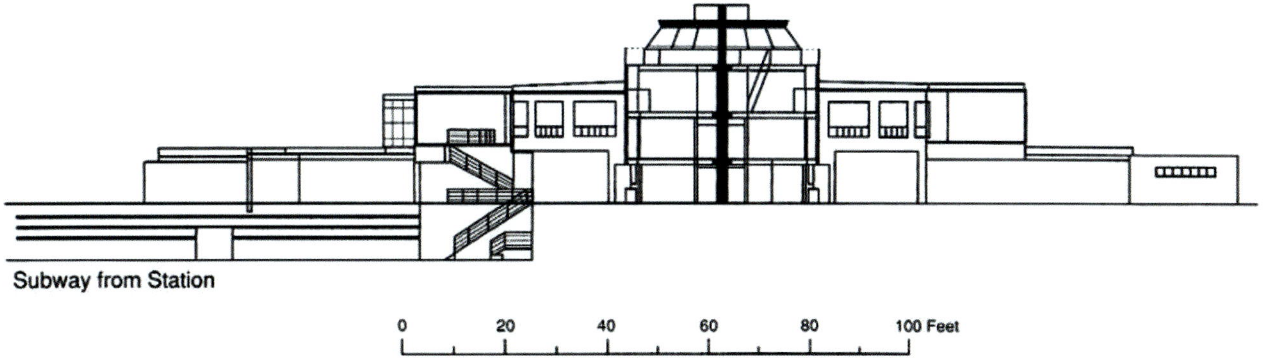

Plan and elevation of the Beehive as it was built, showing the use to which the ground floor parts of the building were put and the subway connecting to the station. The raised centre section housed the airport control tower.
Redrawn from an original in The Architects' Journal.

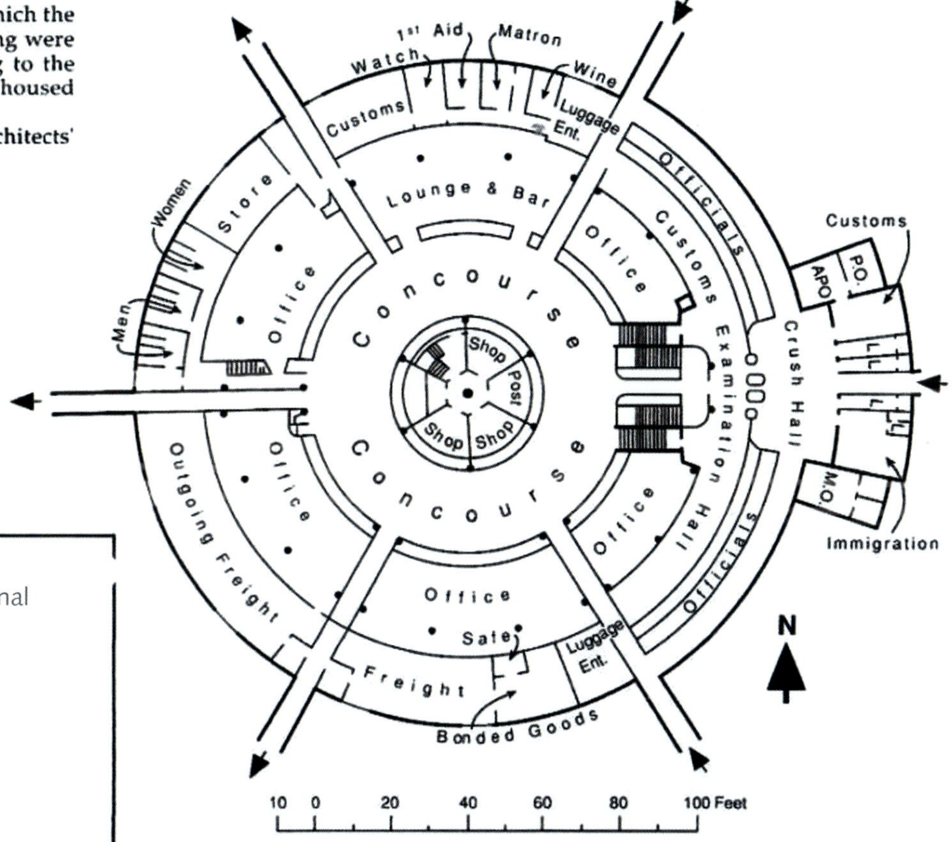

➔ The plan of the Beehive terminal building in 1936.

In 1936, Gatwick became operational as an airport, being officially opened by Lord Swinton, the Secretary of State for Air, on 6 June. The first scheduled flight was by one of four de Havilland DH.86 Express airliners operated by Jersey Airways (not to be confused by the later Jersey Airlines) from Gatwick to Paris. Gatwick had four landing strips made of grass and a subway connecting the Beehive terminal to the train station. The main airline serving Gatwick was British Airways Ltd. Additionally, Southern Railway jointly launched scheduled air services with British Airways between Gatwick and the Isle of Wight.

↓ With the Beehive now completed, here is an aerial view of the airfield in 1936 with the racecourse in the distance.

↑ Up until now, passengers alighting at Gatwick's railway station were going to the racecourse, but the convenience of the railway station beside the airfield was soon capitalised on and the station was renamed Gatwick Airport.

➔ Passengers could now alight from the train straight into the terminal building, as shown on this plan.

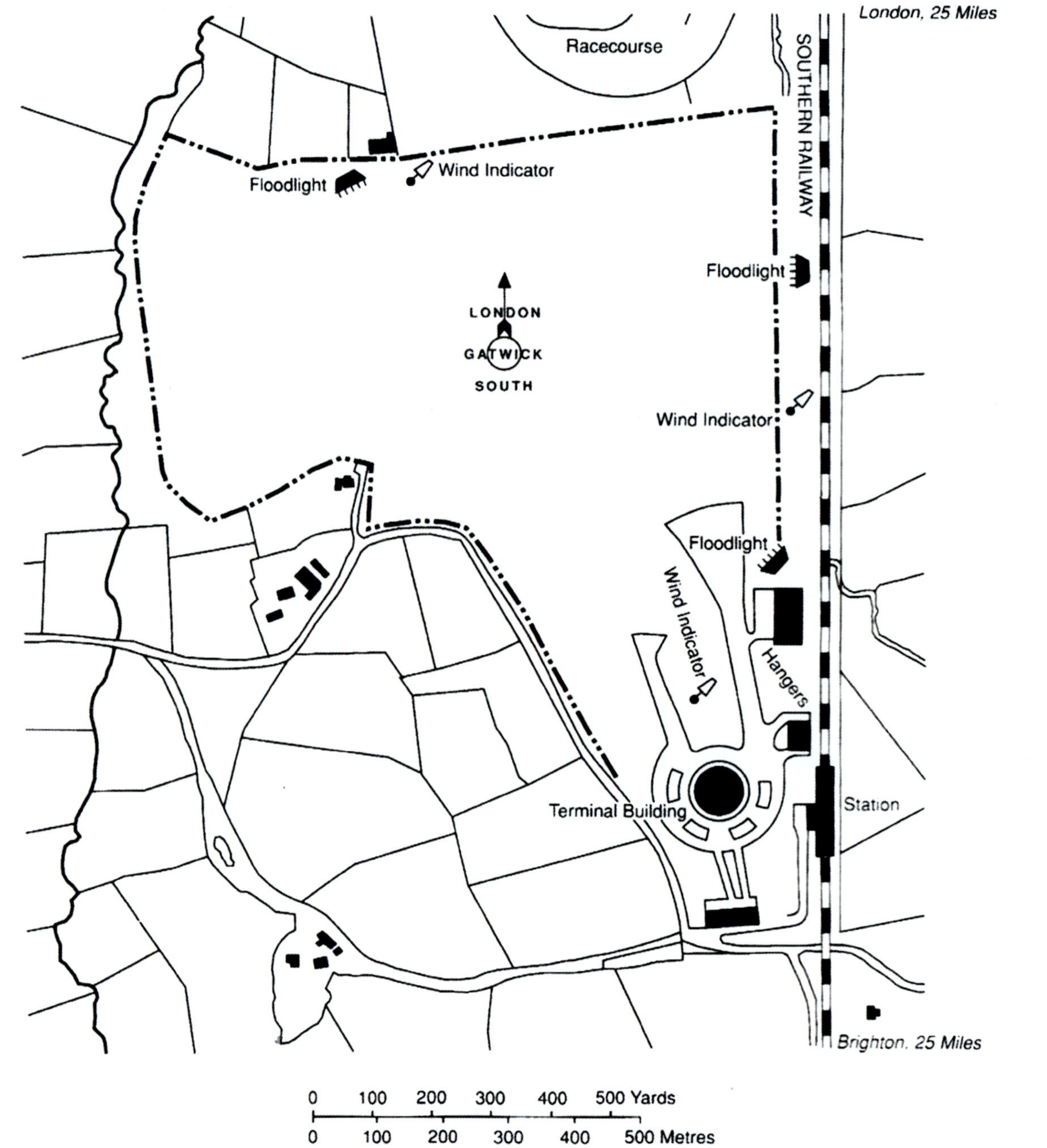

A map of the airport as first developed. The close proximity of the terminal building to the station is very apparent. The title London South, in the centre, was briefly used by Airports Limited while Gravesend was London East but the Air Ministry did not approve.

↑ The Beehive terminal building and control tower was completed in 1936.

← An air traffic controller in the Beehive in late 1930.

↑ 6 June 1936. The UK's then Air Minister, Lord Swinton, arrives in a de Havilland DH.86 Express to officially open Gatwick Airport.

↑ Passengers alighting from British Airways de Havilland DH.86 Express G-ADEA into a purpose-built telescopic walkway into the Beehive terminal building in June 1936.

→ A 1930s airport advert.

← ↓ British Airways Fokker X.II G-ADZK awaits its passengers outside the Beehive terminal in 1936.

An early service was provided by Jersey Airways de Havilland DH.86 Express G-ACZR, competing with British Airways in 1936.

On 25 June 1938, Gatwick hosted the Daily Express Air Display to benefit the RAF Benevolent Fund. Many military and civil aircraft took part, the spectators little realising that this would be the last event here, with dark clouds rising on the horizon in Europe.

When the Second World War broke out in 1939, the airport was requisitioned by the Air Ministry and Gatwick became a base for RAF aircraft including Westland Lysanders, P-51 Mustangs and other types in support of army co-operation and ground-attack duties. The airport then became known as 'RAF Gatwick'. Bomber aircraft were also seen here when servicing for Wellingtons and Liberators were also undertaken here.

At the end of the war, Gatwick's future was uncertain, until the local authorities were persuaded to allow it to continue as a civilian airport for a number of air-taxi and charter companies that had been formed in the immediate post-war years. (The racecourse at Gatwick never fully reopened after the war and today the course has now disappeared below LGW's 1988 North Terminal.)

The first airline to move in was Bond Air Services with an Auster and a Percival Proctor, followed by civilianised Handley Page Halifaxes. That year, the airport re-established a series of air shows, which were very popular. In 1948, Bond Air Services moved out to Southend Airport.

In 1947, BEA set up an Experimental Helicopter Unit at Yeovil to evaluate the potential of the helicopter for regular scheduled operations. They trialled a postal service over four months with Sikorsky S-51s. This was followed by the first passenger helicopter service between Liverpool and Cardiff in 1950. The following year, a helicopter passenger service was started between London and Birmingham, with headquarters in Peterborough.

In April 1952, the Helicopter Unit's HQ was moved from Peterborough to Gatwick, with Sikorsky S-51s and an Agusta-Bell 47J, which became the lasting base of the unit and its successor companies. At first the Helicopter Unit was housed in a small hangar, with a wooden office building, but this was just the beginning of a long residency at Gatwick. The move to Gatwick was to enable scheduled flights from London to be operated more easily and for greater potential for charters in London and the south-east.

↑ Sikorsky S-51 Dragonfly G-AJHW, seen on 16 May 1953, inaugurated BEA's helicopter service.

↑ BEA's Agusta-Bell 47J G-APTH landing at Gatwick in 1959.

TWO

1950S: THE 'NEW' AIRPORT OPENS

More airlines had started to serve Gatwick and among these was Silver City Airways, which since 1949 had already established a very successful car ferry service with Bristol Freighters between Lympne and Le Touquet. In 1953, they purchased three Bristol 170 Mk.32 Superfreighters and received approval to operate vehicle services from other UK airports to Le Touquet, including Gatwick. This commenced in April 1953, with each aircraft carrying three small cars and twenty passengers during the summer months.

A 1952 aerial view of Gatwick showing rows of former RAF Spitfires being refurbished by Airwork for the Burmese and Syrian air forces.

The new terminal building under construction in 1956, showing the close proximity to the railway.

↑ The new car ferry service from Gatwick was started by Silver City Airways in 1953 and here Bristol 170 Mk.32 Superfreighter G-ANWM is seen loading cars in 1958. (Martin Fenner collection)

← 1953 Silver City Airways advert.

With the closure of Northolt as a civil airport on the horizon, it was realised that an alternate to Heathrow would be needed, so BEA continued to press the government for planning consent for a dual runway landing system at Gatwick, which was granted in 1952. Thus, Gatwick was chosen as London's second airport and in 1956 the airport was closed for a £7.8 million renovation, carried out by Alfred McAlpine Ltd.

After two years, it reopened for operations on 30 May 1958, following a major redevelopment project. This new facility, which included the control tower, was a key component of the airport's transformation into a modern, integrated air, road and rail transportation hub. The first aircraft to arrive was Vickers V.804 Viscount G-AOXV, operated by Transair, which arrived from Luqa, Malta, on a trooping flight with military personnel and their families.

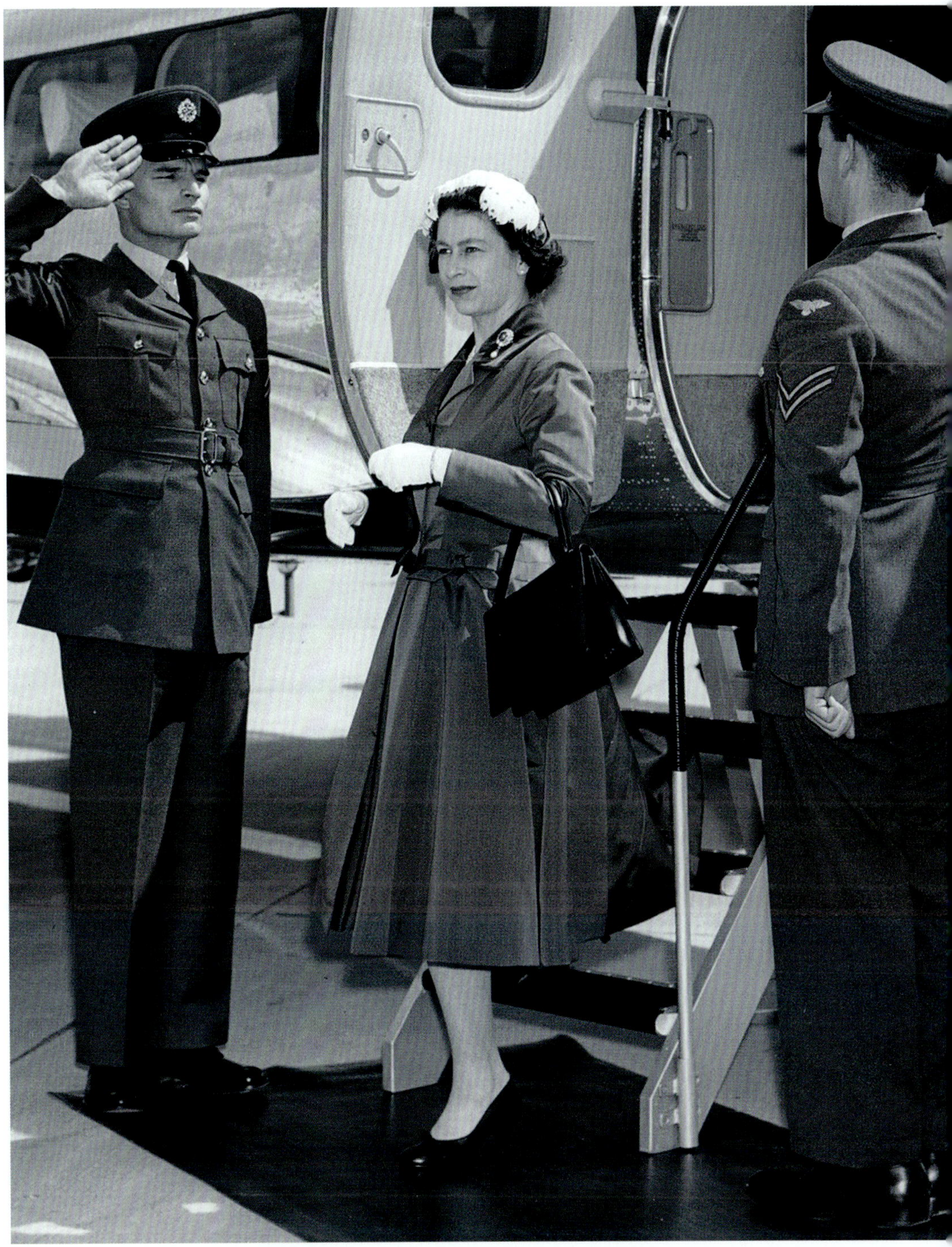

→ ↓ HM the Queen arrives in de Havilland Heron XM296 of the Queen's Flight to officially open Gatwick on 9 June 1958.

↑ HM the Queen arrives in de Havilland Heron XM296 of the Queen's Flight to officially open Gatwick on 9 June 1958.

↑ The first aircraft service to arrive after the official opening was Vickers V.804 Viscount G-AOXV of Transair on 30 May 1958 on a trooping flight from Malta. (Photo via Brian Burrage)

The new Gatwick Airport was officially opened by the Queen on 9 June 1958, Her Majesty arriving in de Havilland Heron C.3 XM296 of the Queen's Flight. Following the official opening, Transair became the first airline to move into Gatwick when it moved in from Croydon. It established itself with a new hangar built that year, together with adjoining offices. A second hangar followed, built by Air Couriers.

→ Transair was the first airline to move into Gatwick in the 1950s and was quickly established with new hangars and offices, following which they took over Airwork.

↓ Air Couriers followed in 1962, building a hangar and offices, and was taken over by Transglobe Airways in 1967. A Caledonian Douglas DC-7C, Vickers Vikings and a Bristol 170 can be seen outside, among other smaller types. (via Brian Burrage)

The first scheduled airline service from Gatwick was a BEA DC-3 charter to Jersey, followed by up to fourteen more flights to Jersey and seven to Guernsey in the same day!

→ ↘ ↓ Interior views of the check-in desks and departure halls in the new South Terminal shortly after opening.

← ↙ ↓ The new air traffic control tower opened in 1958, which replaced the facility in the Beehive.

Gatwick was soon being served by a variety of airlines, including ones from Europe, America, Africa and the Caribbean. More began to join as other airports like Croydon closed.

The layout of the airport included three piers that stretched out on to the apron, alongside which airliners parked. The top of the piers were terraced for spectators and families to 'wave off' their departing loved ones, and plane spotters especially revelled in the new experience, enabling them to get so close to the aircraft. This was a period when the government really started to encourage an interest in aviation and air travel in the UK. Ian Allan recognised this for plane spotters and started publishing an annual book called *Civil Aircraft Markings* and other publications about airports and airliners, where enthusiasts could record what they were seeing. Gatwick was a Mecca for boys interested in trains and planes as both could be seen in close proximity to each other. Identifying the aircraft type, followed by logging the aircraft registration letters, led to identifying the country it came from and thus was a geographical lesson in itself. It is a hobby that many enjoy today, and is much healthier that being sat at home playing video games!

↑ In the 1950s, plane spotting really took off, and with Gatwick's railway station alongside the airport, many train spotters graduated to plane spotting. It is a hobby that continues despite many airports discouraging spectators today.

← The three piers with viewing terraces that extended from the South Terminal were ideal for spectators, as seen in this 1959 view.

← Ian Allan, which had been publishing log books for train spotters, started Civil Aircraft Markings in 1950. These continue to be published and updated annually to the present day. Also in the 1950s, air transport and travel was encouraged with guide books called The Airport Visitor, containing information on the UK's airports and aircraft and airlines likely to be seen.

↓ Many boys didn't need much encouragement to be taken to Gatwick for the day, often to the sufferance of a parent! (Cartoon by Wren from The Airport Visitor)

"Hullo, Mum—where d'you think Dad took us today!"

When Croydon Airport closed in 1959, Morton Air Services moved over to Gatwick with their de Havilland Dragon Rapides, Doves and Herons. Two other airlines started operations at Gatwick that year. Air Links came with Canadair Argonauts, Handley Page Hermes and Douglas DC-3s, while Air Safaris arrived from Southend, also with Hermes and Vickers Vikings, but ceased operations in 1961.

➔ Handley Page HP.81 Hermes G-ALDG of Silver City Airways awaits its passengers in 1961.

➔ Canadair Argonaut G-ALHI being refuelled prior to its next service in June 1965. Also in view are a British United Britannia, Saturn Airways DC-7C, Tunis Air DC-4 and a Spanish TAE DC-6B. (Barry Reed)

← ↓ It's the summer of 1960 and here is a Douglas C-54A G-APCW, one of two operated by Southend-based Continental Air Services, together with a group of happy passengers, all smartly attired, which was the order of the day for this important occasion.

MORTON
AIR SERVICES LTD.
A BRITISH UNITED AIRWAYS ASSOCIATE

TIME TABLE
WINTER 1964/1965
BY AIR
LONDON-SWANSEA-LONDON
MORTON AIR SERVICES LTD.

Avenue 8755 Ext. 295
Crawley 27890 & 296
Gatwick Airport,
Horley, Surrey.

Swansea 24063
Swansea Airport
Swansea

BOOK WITH

↑ Morton Air Services de Havilland Heron Srs.1B G-ANSZ sits awaiting its next task, while a Transglobe Canadair CL-44 is being towed in the distance in September 1968. (Richard Hamblin)

→ Morton Air Services' 1964–65 timetable.

Another Morton Air Services Heron 1B, G-ANWZ, still wearing the colours of Jersey Airlines on 2 September 1967. (Stephen Wolf)

THREE

1960S: EXPANSION FOR AIRCRAFT, PASSENGERS AND SPECTATORS

The Swinging Sixties with Props and Jets Galore

In the 1960s, it was expensive to fly from Heathrow on the major flag-carrying airlines. If you wanted a family package holiday, the chances are you flew from Gatwick, Luton, Manchester or Birmingham.

As Gatwick became operational, two airlines dominated from the start. In July 1960, Hunting-Clan Air Transport and Airwork merged to form British United Airways (BUA), becoming the UK's largest private airline at the time. The airline inherited a fleet of Douglas DC-3s, DC-6s, Bristol Britannias and Vickers Viscounts, but were looking for new jet airliner equipment.

↑ Wearing best hats and high heels, this looks like a Women's Institute charter boarding Vickers V.708 Viscount G-ARGR of British United Airways in 1965.

↑ When British United Airways absorbed several smaller airlines, not only were the aircraft repainted in their livery, but also all their ground support vehicles, including these two Commer vans. (via Tom Singfield)

→ British United Airways ticket c.1965.

↑ British United Airways Bristol Britannia Srs.307 G-ANCE alongside one of the piers in 1965. (John Mounce collection)

↑ Following the formation of British United Airways and the acquisition of new equipment, some aircraft were relegated to other duties including this DC-3, G-AOBN, which was used by the airline's Radio Calibration Unit at Gatwick in 1965.

British United's reservations desk in the South Terminal's departures hall in the early 1960s.

A 1963 British United Airways magazine advertisement.

→ ↓ Following the takeover of Silver City Airways by British United, occasionally their Bristol Freighters were replaced by Aviation Traders ATL-98 Carvairs, but they were rare visitors to Gatwick. Here British United Air Ferries Carvair G-ARSD taxies into position, dwarfing the private aircraft parked behind it. (Barry Reed)

The second airline, which also moved to Gatwick in 1960, was Dan-Air, a subsidiary of London-based shipbroking firm Davies and Newman. It was started in 1953 with a single DC-3 aircraft, operating cargo and passenger charter flights from Southend (1953–55) and Blackbushe airports (1955–60) using a variety of piston-engined aircraft, before expanding into inclusive tour charter flights and all-year round scheduled services. When moving to Gatwick, it initially operated Airspeed Ambassadors, Avro Yorks, Douglas DC-3s and Bristol Freighters.

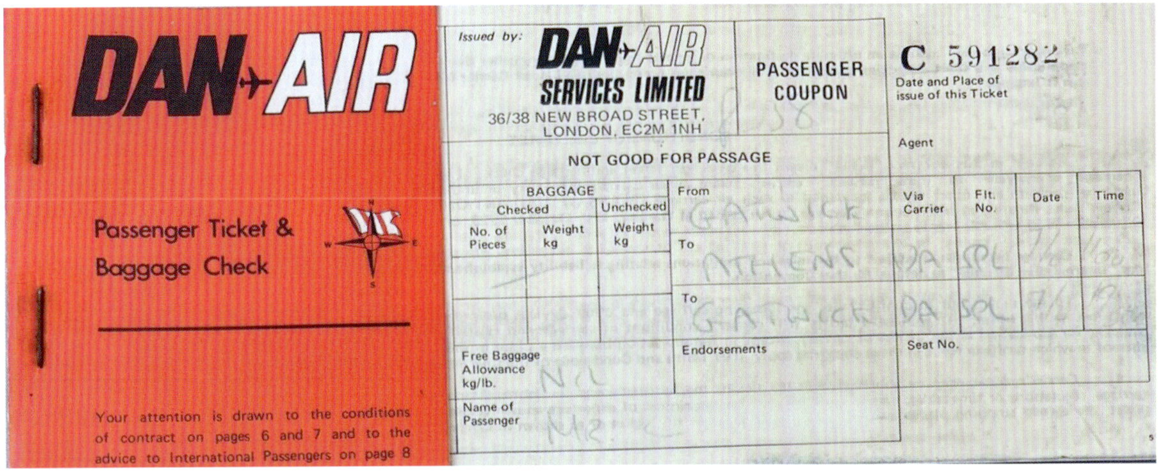
DAN-AIR

Passenger Ticket & Baggage Check

Your attention is drawn to the conditions of contract on pages 6 and 7 and to the advice to International Passengers on page 8

Issued by: DAN-AIR SERVICES LIMITED
36/38 NEW BROAD STREET, LONDON, EC2M 1NH

PASSENGER COUPON

C 591282

Date and Place of issue of this Ticket

Agent

NOT GOOD FOR PASSAGE

BAGGAGE — Checked: No. of Pieces, Weight kg; Unchecked: Weight kg

From / To / To

Via Carrier | Flt. No. | Date | Time

Free Baggage Allowance kg/lb.

Endorsements

Seat No.

Name of Passenger

➔ 1960s Dan-Air passenger ticket.

🡑 Dan-Air operated a fleet of seven Airspeed Ambassadors from 1960 for a decade until replaced by Comets. (R.G.Griggs via Stephen Wolf)

🡑 A rare flying shot of Dan-Air's Avro York G-ANTK as it comes into land at Gatwick on 24 March 1964. This aircraft is now preserved at Duxford. (Steve Bond)

↑ Dan-Air's Bristol 170 Mk.31 Freighter G-APLH sits at Gatwick on 19 June 1965 awaiting its next task. (Clive Moggridge)

In September 1961, Silver City Airways moved its fleet of five Handley Page Hermes to Gatwick from Manston to carrying out a contract of trooping flights to bases in Germany, as well as to fly inclusive charter flights to destinations in Europe and the Mediterranean. In 1962, Silver City was taken over by British United Airways and the Hermes fleet was withdrawn from use, while the Bristol Freighter fleet formed British United Air Ferries.

At the same time another airline was formed that would end up being a major carrier operating from Gatwick for years to come. This was Caledonian Airways, which formed on 27 April 1961 with two Douglas DC-6Bs leased from Belgian Airline SABENA, before acquiring a fleet of three Douglas DC-7Cs. Caledonian was a Scottish international flag carrier, with most of their funding coming from Scottish businesses, that was keen to focus on the large emigrant communities in North America.

In 1963, the American CAB granted a licence to Caledonian as the first UK charter operator to fly the transatlantic route to the States from Gatwick. Caledonian promptly acquired a further five DC-7Cs and the Ministry of Aviation transferred all regular charter flights from Heathrow to Gatwick.

↑ Caledonian's Douglas DC-7C G-AOIE is seen inbound to Gatwick on 31 July 1965. (Steve Bond)

← 1960s Caledonian advertisement introducing the Britannia.

↑ Caledonian's Bristol Britannia Srs.314 G-ASTF departs on schedule, with its 'cousin', an RCAF Canadair Yukon, behind on 21 April 1968.

↑ Caledonian's first jet equipment, Boeing 707-339C G-AVTW, at Brisbane in December 1970. (John Roger collection)

For the summer of 1965, Caledonian started replacing its DC-7C fleet with the first of nine turboprop Bristol Britannias mainly purchased from BOAC and Cunard Eagle. That year, Caledonian Airways was well established and in a position to contemplate the purchase of bigger and more modern aircraft. In November, an order was placed for a Boeing 707-320C to be delivered in May 1967. In July 1966, a second 707-320C was ordered for 1968 delivery. Meanwhile, by 1967, the Britannias had completely replaced Caledonian's DC-7C fleet.

BEA still had its Helicopter Unit based at Gatwick and was now operating Westland-built Sikorsky S-55 Whirlwind helicopters on charter contract work. By 1961, it had graduated to a modern building incorporating both an engineering base and administrative offices located on the south side of the airport near the Beehive terminal building.

In 1963, the BEA Helicopter Unit received some publicity and in particular Westland Whirlwind G-ANFH, which was featured in the Beatles' film *A Hard Day's Night*. In the movie, it's depicted carrying the band away from a crowd at the end of the film, with the BEA logo on the fuselage altered to also read 'TLES'. It also appeared in several other films, including a starring turn in the 1965 Terry Thomas comedy *You Must Be Joking*. In July 1966, it was flown to Brands Hatch to be an air ambulance for the British Grand Prix. It was acquired by the Helicopter Museum at Weston-Super-Mare in 1978, where it is currently stored awaiting restoration.

↑ BEA's Westland S-55 G-ANFH operated from Gatwick in 1963 when being used in the Beatles' film A Hard Day's Night.

↑ New equipment for BEA's Helicopter Unit arrived in 1963 in the form of the first two Sikorsky S-61Ns.

In January 1964, two brand-new Sikorsky S-61N helicopters arrived for BEA. While these would be maintained at Gatwick, they were for BEA's new route between Penzance and the Scilly Isles, which turned out to be very successful. The following year, BEA decided to move its Helicopter Unit out of Gatwick to Beccles Airfield, in order to support the North Sea oil and gas industry with Sikorsky S-61s.

In 1964, the runway was extended to 2,500m long. By the next year, the airport had three piers nearly 300m long and a terminal floor area of 9,300 sq m. Gatwick was the first airport in Britain to feature a 'finger' terminal, along which passengers would walk to their aircraft parked on either side. This system was adopted by most of the major airports of the world and was considered more efficient than the channel and bus system used for the short-haul passengers at Heathrow. Initially a single central pier was constructed, but to accommodate the increasing number of travellers using the airport, two additional piers were added to form a huge 'trident'.

↘ Plan of Gatwick in the 1960s.

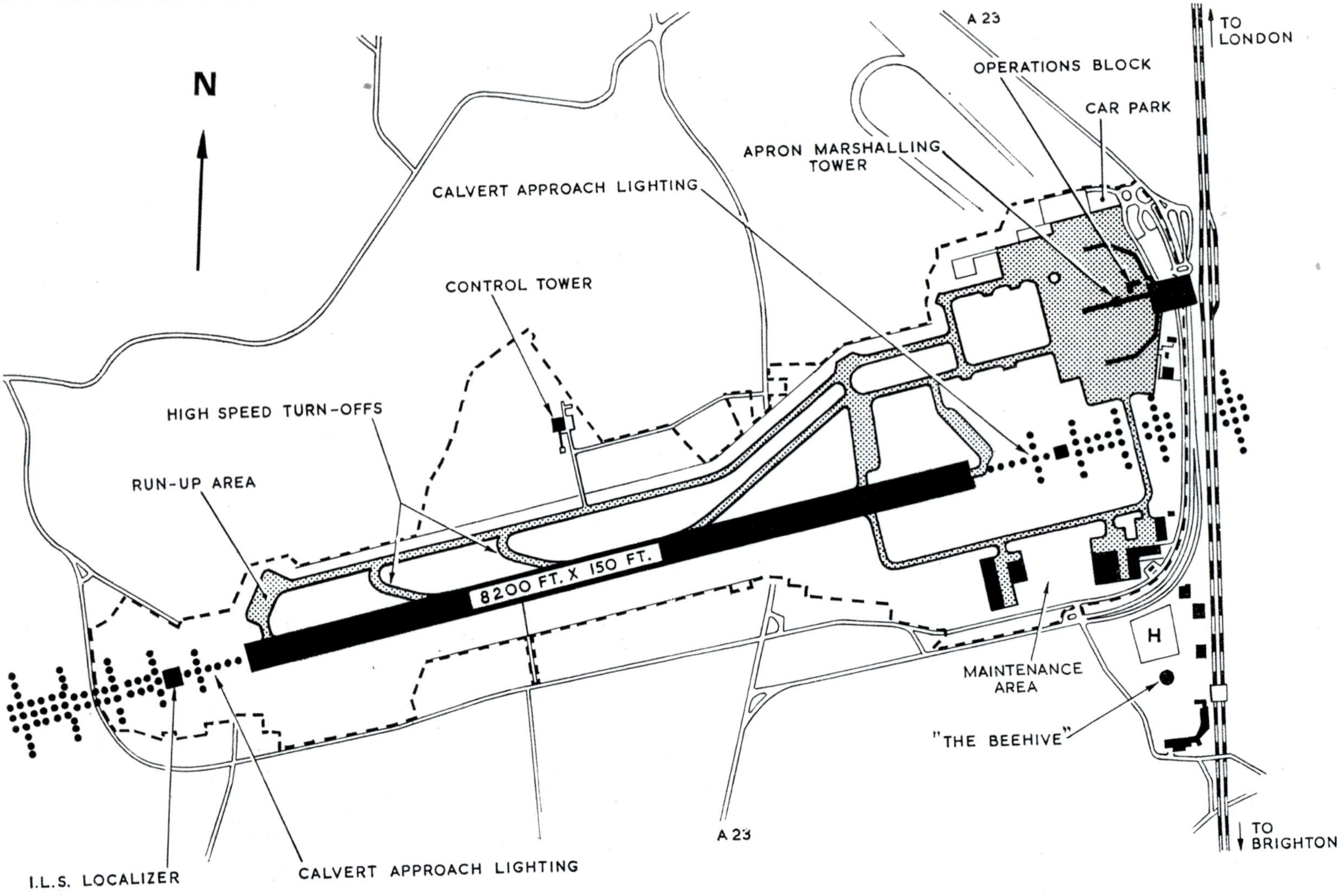

↑ Luxair was the last airline to operate the Lockheed L-1649A Starliner into Gatwick and LX-LGY is seen here arriving from Luxembourg on 15 August 1968. This aircraft was withdrawn from use the following year. (Stephen Wolf)

Spectators' roof terraces were provided on the roofs of these piers with a café and souvenir shop, giving visitors a magnificent view of the activity on the surrounding aprons. The 1960s was a haven for plane spotters at Gatwick, which now had a diverse variety of aircraft from vintage piston-engined and turboprop airliners to the latest modern jets, operating scheduled flights and charters.

It was even better than Heathrow, with aircraft parked close up on either side of the piers. Telephoto lenses on cameras were a new expensive luxury in those days, but were completely unnecessary as the aircraft parked so close, often too close for the average camera! However, when the airport was first opened and spectators were first allowed onto the central pier, their general behaviour gave cause for concern. The throwing of empty cigarette packets and other litter onto the apron was extremely disrespectful, thoughtless and irresponsible, embarrassing apron staff and passengers alike. The throwing down of cigarettes was positively dangerous, with aircraft refuelling within close proximity! Safety and security were the reasons these spectators' piers were finally closed to the public, restricting them to the roof of the main terminal building.

↑ This is October 1964 on Gatwick's wonderful spectators' viewing piers – boys dressed up smartly for a plane spotting trip, but trying to impress their girlfriend with their hobby on a first date could be tricky! (John Kimberly via Tom Singfield)

➔ The spectators' viewing terraces were spectacularly close to all the action, as seen by this Saudi Arabian Airlines Douglas DC-6B …

➔ … also by this Austrian Airlines Vickers V.837 Viscount, OE-IAM, both on 15 August 1968 (Stephen Wolf)

A rare gem for plane spotters in 1968 was this Douglas DC-6B, YA-DAO of Ariana Afghan Airlines. On the right the South Terminal's expansion is seen under construction. (John Mounce collection)

The first of British United Airways' new jet airliners arrived with the first of two new VC10s. G-ASIW was delivered on 30 September 1964 and immediately began operating on Middle Eastern and African trooping services. The second example, G-ASIX, arrived a month later and was introduced on scheduled routes to Africa, including Freetown, Entebbe, Nairobi, Ndola, Lusaka and Salisbury (now Harare). It was even leased to Sierra Leone Airways for a short period.

↑ A 1964 British United Airways advertisement.

↑ British United Airways' brand-new jet equipment was delivered in 1964, including two Vickers VC10s. G-ASIX is seen here in June 1966. (Barry Reed collection)

↑ British United was the first customer for the BAC One-Eleven and here G-ASJJ awaits passengers for its inaugural flight to Genoa on 9 April 1965. Note British United Handley Page Herald G-APWH in the background. (Barry Reed collection)

British United Airways (BUA) was the launch customer for the BAC One-Eleven in the UK, receiving its first aircraft in early April 1965. BUA placed an initial order for ten in May 1961, marking the start of its fleet re-equipment programme as it transitioned to a scheduled airline.

The world's first commercial flight of a One-Eleven took place in April 1965 when British United's G-ASJI departed Gatwick for Genoa. While BUA primarily used the early short-fuselage versions, BUA later ordered the longer 500 series variant, which could carry more passengers.

In 1967, BUA adopted a new company livery image in a sand and light blue scheme, probably representing two major holiday selling points: blue skies and a sandy beach. This also perpetuated throughout the company, extending to all ground support equipment and new cabin staff uniforms.

→ In 1967, BUA changed the livery on its aircraft to this attractive light blue and sand brown scheme. BAC One-Eleven 201AC G-ASJD is on stand in June 1967 wearing the new colours with matching ground support vehicles. (Stephen Wolf collection)

↓ BUA's two VC10s were also repainted in the new livery, with G-ASIX seen in October 1969. (Stephen Wolf collection)

↑ A 1967 BUA advertisement and 1968–69 timetable promoting the new livery and air hostess's uniforms … and just look at the prices!

↘ British United's Bristol Britannia 317 G-APNB arrives back from a holiday destination on 21 April 1968.

➔ A 1967 BUA passenger ticket.

↓ British United Handley Page Herald G-APWH departs on 11 May 1968. The South Terminal building's expansion has now been completed. (Stephen Wolf)

One short-lived Gatwick-based airline was Transglobe Airways, which was founded in 1965 after changing its name from Air Links. It operated holiday and cargo services from Gatwick initially with three Bristol Britannias. In May 1967, it had an agreement with Seaboard World Airlines for the lease-purchase of six Canadair CL-44 swing-tail freighters, five of which were the convertible passenger/cargo model while the sixth was pure cargo. The first arrived in April 1968, followed by two more, but the rest were never delivered due to the airline ceasing operations on 28 November that year.

→ Transglobe Airways' business card.

↓ Transglobe's Bristol Britannia 314 G-ATLE departs on another charter flight on 21 April 1968.

↑ In 1968 Transglobe's fleet were given a new livery, as seen here on Bristol Britannia 302 G-ANCC on 24 November. It was short-lived, as this aircraft was withdrawn from use soon after this photo was taken. (Stephen Wolf)

➔ The new livery was also implemented on its leased Canadair CL-44s, including G-AWDK, seen on 15 August 1968. (Stephen Wolf)

➔ When Transglobe ceased operations in November 1968, it was revived as Tradewinds in January 1969 and continued to receive its full complement of eight leased Canadair CL-44s, including G-AWGS seen in October 1970.

However, like a phoenix out of the ashes, the airline re-emerged as Tradewinds Airways Ltd in January 1969 and continued to receive the full complement of CL-44D4s from Seaboard World Airlines. During the civil war with Biafra, Tradewinds flew several flights to Nigeria, allowing the company's expansion into Africa, the Middle East and the Far East. They also flew several flights under contract to the UK Ministry of Defence. Interesting loads included transporting missiles to the NATO Arctic test ranges as well as an exclusive contract for transporting Formula One racing cars. Tradewinds became a specialist carrier for bloodstock, as well as allowing many champion horses to travel safely and without stress in specially designed stalls. Not only horses, but sheep, goats and pigs were carried among other livestock. The airline's base at Gatwick included its own warehouse, a privilege only shared by British Airways and British Caledonian at the time. Other airlines had to use either Gatwick Handling services or another carrier for handling.

Despite the Cold War in the 1960s, the Soviet Union and Eastern Bloc countries were still trying to promote themselves as friendly holiday destinations. Aeroflot was operating Tupolev Tu-104s and Ilyushin Il-18s into both Heathrow and Gatwick Airports once a week.

➔ Despite the Cold War being at its height in the 1960s, the Soviet Union still tried to attract tourism with this advert.

↑ Aeroflot Tupolev Tu-104s were regular Gatwick visitors in the 1960s, which continued until the early '70s.

→ A rare 1960s Soviet Aeroflot ticket from Gatwick to Moscow.

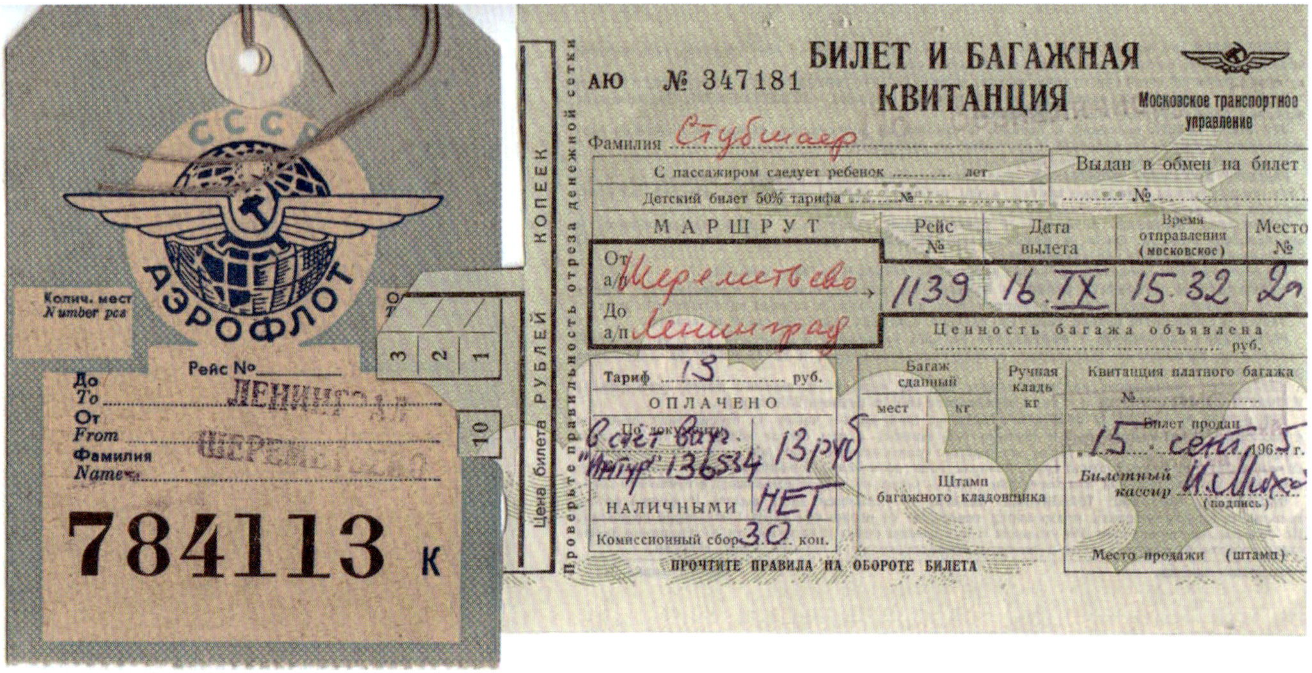

On 14 March 1967, a Soviet Mil Mi-10 'Harke' helicopter arrived on a sales demonstration flight at Gatwick, carrying a 30ft coach underneath. At the time, it was the world's largest helicopter and could carry loads up to 25 tons, with an estimated cruising speed of 130mph and a range of some 150 miles. European buyers might have been put off by the £600,000 price tag!

↓ Another East Bloc country promoting tourism was Romania, whose Ilyushin Il-18s were seen regularly at Gatwick. YR-IMF of Tarom is seen in July 1965 with a DC-6B of the French charter airline UTA (Union de Transports Aériens) in the background. (via Barry Reed)

➔ ➘ The Russian Mil Mi-10 'Harke' helicopter at Gatwick in March 1967 on a European sales tour.

Meanwhile, BUA managing director Freddie Laker had become increasingly frustrated with his board and yearned to be his own boss once again, so he resigned in November 1965. In February 1966, at the age of 43, he launched Laker Airways as a low-cost airline, dealing with inclusive tours (IT) for the holiday market.

Development of the BAC One-Eleven continued despite the loss of two aircraft during the flight test programme. As launch customers, Laker and BUA waited patiently, while in late 1964 and early 1965 the type underwent proving trials on the airline's network, including the Mediterranean and West Africa.

With his close ties with BAC, Laker was able to gain sufficient financial backing to order three One-Eleven 300s.

G-AVBX, the second of two newly delivered BAC One-Elevens for Laker Airways, in May 1967. Note the Southern Terminal expansion under construction. (Stephen Wolf collection)

When Laker Airways was formed in July 1966, their first aircraft were two former BOAC Bristol Britannia Srs.102s, including G-ANBN seen here on 2 September 1968. (Stephen Wolf)

He also took delivery of a former BOAC Britannia in April and, after an overhaul by Scottish Aviation at Prestwick, G-ANBM arrived at Gatwick in the now-familiar red-and-black-striped fuselage scheme with Laker titles on the fin.

The aircraft was wet leased to Air France for six weeks and used on scheduled services to Paris. For the rest of the period it was used on ad hoc and freight charters.

Laker Airways took delivery of its first One-Eleven, G-AVBW (c/n 107), on 25 February 1967 with commercial operations beginning in March with a series of IT charter flights from Gatwick and Manchester to Mediterranean resorts on behalf of Lord Brothers (Travel).

The second example, G-AVBX (c/n 109), was delivered on 8 April and a third, G-AVBY (c/n 113), was wet leased to Air Congo, departing from Hurn to Kinshasa, Democratic Republic of Congo, on 9 May for twelve months for use on scheduled services.

Laker acquired Liverpool-based tour company, Arrowsmith Holidays, enabling his One-Elevens to operate inclusive tour charters from the north-western city. He also introduced time charters that tied tour companies into contracts, which guaranteed his airline a fixed number of flying hours each year to carry their customers to popular Mediterranean sun spots. This arrangement allowed him to sell seats on the One-Elevens up to two years in advance.

Meanwhile, with the One-Elevens now in service, during the summer months, Laker leased Britannia G-ANBM to short-lived Treffield International and G-ANBN was leased to Luton-based Britannia Airways to augment its flying programme.

For the 1968 season, a fourth One-Eleven, G-AVYZ (c/n 133), joined the fleet, undertaking IT charter flights on behalf of leading tour operators, including Arrowsmith Holidays, Lord Brothers and Wings. The arrival of the last example spelt the end for Laker's Britannia fleet, and both aircraft were sold in Indonesia in January 1969.

As a result of the restrictions placed on West German carriers after the Second World War, Laker saw the opportunity to establish a One-Eleven base in what was then West Berlin, operating IT services to Mediterranean destinations. Over time, this proved to be a fruitful plan and it developed into an intercontinental hub for the airline.

↑ Laker's two Britannias were later leased, G-ANBN to Britannia Airways, as seen on 21 April 1968.

↑ Laker's Britannia 102 G-ANBM was leased to short-lived start-up airline Treffield International in May 1967.

To supplement its aging fleet of Airspeed Ambassadors and other piston-engined types, in 1966 Dan-Air moved into the jet age, acquiring its first former BOAC de Havilland Comet 4. This was the start of an affection for the Comet and as BOAC, BEA and other overseas operators disposed of their Comets, Dan-Air acquired them all, and they were refurbished at their engineering base at nearby Lasham. In August 1975, when the RAF disposed of their five Comet 4Cs, Dan-Air also acquired those, which, with low flying hours, were in excellent condition. Dan-Air was the world's largest and last operator of the Comet, building a fleet of forty-nine aircraft between 1966 and 1976.

➔ Dan-Air Airspeed Ambassador G-AMAH receives attention to one of its engines in 1968. The airline operated ten Ambassadors, which were its first pressurised aircraft and were instrumental in expanding its IT charter market in the mid-1960s. This was one of the last in service and it was withdrawn from use in October 1970.

➘ G-APDD was one of the first Comet 4s purchased from BOAC by Dan-Air and is seen here in May 1967. The Comet was Dan-Air's first jet equipment, which they loved, going on to acquire the world's remaining airworthy examples. (Stephen Wolf collection)

Already mentioned, Morton Air Services, a British operator, was one of the first post-Second World War private, independent British airlines. Formed in 1945 and operating from Croydon Airport, its main operations were within the British Isles on regional short-haul scheduled services from and to the United Kingdom and Continental Europe. Olley Air Service was a rival independent airline, so Morton took them over in 1953 before, in 1958, being absorbed itself into the Airwork group, which operated DC-3s, Doves and Herons, moving from Croydon to Gatwick. Following the Airwork-Hunting Clan merger which created British United Airways in 1960, Morton retained its identity. However, when the BUA group of companies were reorganised during 1967–68, Morton was absorbed into British United Island Airways (BUIA) in 1968.

↑ Morton Air Services' 1967 timetable.

← Morton Air Services' fleet of DC-3s and Herons were a regular part of the 1960s scene and DC-3 G-AMYJ is seen here on 11 May 1968 departing on another service. (Stephen Wolf)

← Morton Air Services' passenger ticket.

↓ When Morton Air Services was taken over by British United Airways, their de Havilland Herons became part of the British United Island Airways fleet. Heron Srs.1B G-ANSZ is seen in its new BUIA livery in October 1969.

As the major flag carriers over at Heathrow re-equipped their fleets with more modern airliners, their older equipment was often sold to start-up airlines and charter carriers, which could now be seen at Gatwick. Military transport aircraft also used Gatwick, mainly Royal Canadian Air Force Canadair Yukons (military CL-44s), Dakotas, C-130 Hercules and Bristol Freighters on personnel and freight flights shuttling between the RCAF bases in France and Germany. At Gatwick, the Canadian military personnel would then connect with charter flights by Wardair Boeing 707s or Canadian Pacific DC-8s. US military aircraft would also be seen at Gatwick on liaison flights from their bases in Europe.

↑ Canadian military aircraft were regular visitors, including Douglas C-47 KP221 of the RCAF's 1 Air Division from Marville, France, on 21 November 1968. Note the variety of military and civil aircraft in the background! (Stephen Wolf)

↑ The variety of aircraft seen at Gatwick included military transports commuting from their European bases, including US Army Beechcraft U-8D Seminole 57-3087 on 21 April 1968.

↑ RCAF Bristol 170 Mk.31M Freighter 9698, seen in 1965, was one of a handful that transported freight between Gatwick and Canadian bases in France and Germany. (John Mounce collection)

← The Bristol Freighters were later replaced by RCAF Lockheed Hercules, including C-130E 10320, seen on 22 July 1967.

↓ Canadair licence-built the Bristol Britannia as the CL-44 with Rolls-Royce Tyne engines and other improvements. The RCAF operated them as the CC-106 Yukon, as seen here by 15923 on 10 July 1969, transporting Canadian military personnel between Gatwick and Europe.

→ A handful of Canadian Armed Forces DHC-5 Buffalos were operated in Europe, but these were rare visitors to Gatwick. Here 9259 is seen on a dismal, wet day on 22 June 1968. Also of note is that the South Terminal building expansion has now been completed.

↓ Canadian charter airline Wardair was contracted to bring Canadian military personnel and their families to and from Canada, connecting with the RCAF transport aircraft at Gatwick to take them to their onward destinations in Europe. Boeing 707-311C CF-FAN is seen here on 11 May 1968 in company with a Capitol Airlines DC-8. (Stephen Wolf)

↑ Another rare military visitor on 18 August 1968 was Fairchild C-123B Provider 54505 of the US Coast Guard based at Naples, Italy. (Jeff Peck)

Under the chairmanship of Peter Masefield, the British Airports Authority was formed on 1 April 1966, transferring control from the Ministry of Aviation. On the same day a new General Aviation terminal was opened, these aircraft accounting for 20 per cent of aircraft movements.

→ When the general aviation terminal was opened, it was well used by a variety of executive and private aircraft including Spanish Beagle B.206S EC-BJF (Ex G-ATYX) of Aviones Taxis, seen on 21 April 1968.

↑ Larger executive aircraft such as this Allison-turboprop powered Convair 580 N270E of Engelhard Industries was one of the more interesting visitors in May 1966. (Barry Reed collection)

↑ The Dutch Philips company fleet were regular visitors during the 1960s, the largest of which was this Fokker F.27-100 Friendship, PH-LIP, seen on 24 November 1968. (Stephen Wolf)

On 1 October 1967, British United Air Ferries changed its name to British Air Ferries (BAF) and became a completely separate division from its parent British United Airways, but with competition from sea ferries, their Bristol Superfreighters were now used more for cargo and passengers than ferrying cars across the English Channel. The Superfreighters were still occasionally seen at Gatwick on cargo charter flights, but were finally withdrawn from use at the end of 1970.

↑ Up until their retirement, the Bristol 170 Mk.32 Superfreighters of British Air Ferries were still occasional visitors. Here G-APAV is seen in October 1969, clearly its last visit as it was withdrawn from service soon after.

In March 1968, Caledonian decided to start replacing its Bristol Britannias with a fleet of One-Elevens and initially ordered three Series 500s.

↑ G-AWWZ was one of four Caledonian Airways' initial fleet of BAC One-Eleven 500s, seen on October 1969. (Stephen Wolf collection)

All went well at Gatwick until 5 January 1969, when a Boeing 727 operating Afghan Airlines' Flight 701 was involved in an accident with sixty-two passengers on board, crashing into a house during its approach amidst heavy fog. The main cause of the accident was down to pilot error, specifically when the flaps were not extended in order to maintain flight during the final approach speed.*

* AAIB Accident Report No. EW/C/303.

→ ↓ Tragedy struck the airport when Ariana Afghan Airlines Boeing 727-113 YA-FAR crashed on approach in fog. (Stephen Wolf and author's collection)

It was 1:35 on a Sunday morning that the accident occurred, during which the Gatwick area was enveloped in patches of dense, freezing fog. YA-FAR, the sole Boeing 727 in the fleet, operated by the airline at the time, approached the airport from the east, descending below its correct glide slope. As the aircraft passed over the small hamlet of Fernhill, located on the Surrey/Sussex border, it was too low and struck trees and rooftops, resulting in a roll. It subsequently crashed into a field south of Fernhill Lane, which was approximately 1.5 miles (2.4km) short of the runway. Following the initial impact, the aircraft then collided with a large detached house, resulting in its complete demolition and bursting into flames. The flight's captain, first officer, flight engineer, and eleven passengers survived the crash, but forty-eight passengers and crew members perished. In addition, two adult residents in the demolished house lost their lives due to the impact. Amazingly, there was an infant present in the house who survived the incident with only minor injuries.

A great variety of vintage piston-engined airliners from airlines worldwide could be seen at Gatwick throughout the 1960s, the most numerous among these being the Douglas DC-6B and DC-7C.

↓ A murky wet day did not dampen the author's excitement when he saw Ethiopian Airlines Douglas DC-6B ET-AAX sharing the apron with Dan-Air's DC-6B G-ATAB on 22 June 1968.

→ Ethiopian Airlines DC-6 passenger ticket.

↑ The 1960s were a golden age for plane spotters and the Douglas propliners were very much part of the scene, including regular visits by the DC-6Bs of SAM (Società Aerea Mediterranea), including I-DIMP on 15 August 1968. (Stephen Wolf)

↑ Pacific Western Airlines Douglas DC-7C CF-NAI being refuelled in June 1966, probably a charter flight for Canadian service personnel. (Barry Reed collection)

↑ Swedish airline Transair operated charter flights with a fleet of former Eastern Airlines Douglas DC-7Bs, including SE-ERE, seen arriving on 15 August 1968. (Stephen Wolf)

← Transair also had a fleet of eleven Curtiss C-46 Commandos for passenger and cargo work, including SE-CFA, seen in May 1965. (Barry Reed collection)

↑ Kar-Air of Finland operated a pair of passenger-carrying Douglas DC-6Bs on scheduled and charter flights including OH-KDB, seen in June 1966, while a third was converted to a swing-tail freighter. (Barry Reed collection)

↑ Seen on 28 October 1969, Swissair Convair 440 HB-IMM was re-engined with Rolls-Royce Dart turboprop engines, emerging as a Convair 640 to be operated by Swiss charter airline SATA.

↑ An early 1960s aerial view of Gatwick. (Gatwick Airport Authority)

FOUR

1970S: THE RISE AND FALL OF AIRLINES

In 1968, British United Airways was running into financial difficulties, despite during the summer of 1970 using its stretched One-Eleven on a greatly expanded route network to holiday destinations such as Dubrovnik, Faro, Tenerife and Zurich. So on 20 October 1970, an agreement was reached whereby Caledonian Airways would take over British United Airways to form Britain's second largest airline. Initially the merged airline was known as Caledonian/BUA but the name was eventually changed to British Caledonian Airways.

↑ When British United Airways merged with Caledonian Airways, the aircraft initially wore a hybrid livery with Caledonian/BUA titles, as on BAC One-Eleven 201 G-ASJI in May 1971, seen parked alongside Wardair Boeing 727 CF-FUN. (Stephen Wolf)

↑ British United's Douglas C-47s, including G-AMHJ seen in April 1971, were transferred to the British United Island Airways division and converted to freighters. (Werner Gysin-Aegerter)

→ By November 1971, most of the merged BUA/Caledonian fleet had been repainted as British Caledonian as seen on BAC One-Eleven 509EW G-AXJL, alongside another still in the hybrid livery. (Stephen Wolf)

↓ One interesting anomaly of the BUA/Caledonian merger was Boeing 707-323C G-AYZZ, which was formerly N8417, being leased from American Airlines in August 1971, and wore this mixed livery.

← Early 1970s BCAL advertisement.

↓ A pair of British Caledonian BAC One-Elevens and a Boeing 707 on 7 July 1975. (Stephen Wolf collection)

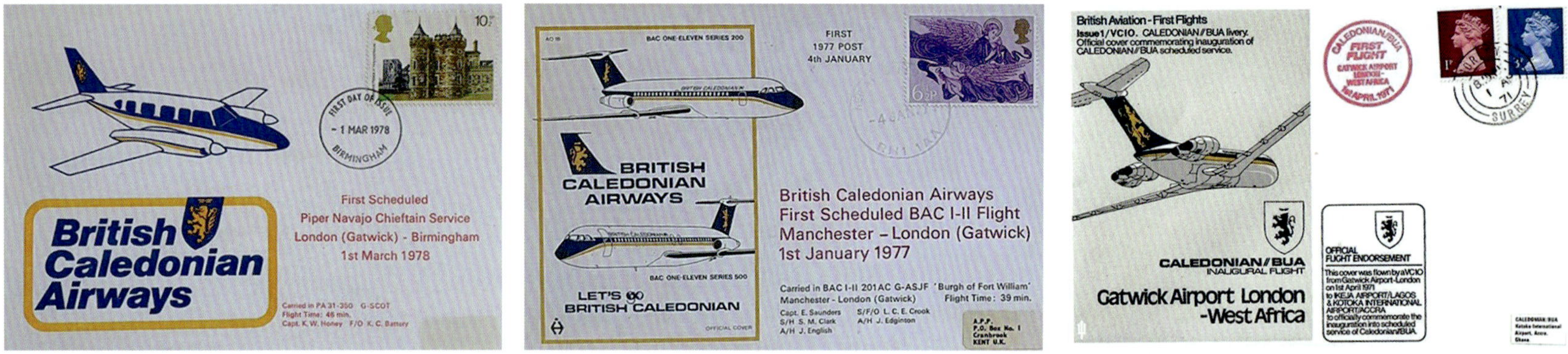

↑ BCAL first day flight cover envelopes.

↓ British Caledonian's Gatwick base with a VC10 outside in June 1973. In the distance can be seen the original Beehive terminal. (Tom Singfield collection)

In 1970, the runway was further extended to 2,766m in length, enabling heavily laden jet flights to depart from Gatwick bound for the US East Coast.

That year, BEA formed a subsidiary division called BEA Airtours, transferring nine Comet 4Bs over from Heathrow to Gatwick. By the end of the year, this was so successful that they were looking at expanding the fleet and two years later acquired the first of ten Rolls-Royce Conway-powered former BOAC Boeing 707s. By the end of the 1973 season, only five Comets remained in service and these were finally withdrawn at the end of the year. When BEA and BOAC merged on 1 April 1974 to form British Airways, BEA Airtours was renamed British Airtours.

↑ By November 1971, BEA had transferred part of its Comet 4B fleet from Heathrow to Gatwick to form BEA Airtours' charter division, including G-ARJK. (Stephen Wolf).

↑ A mid-'70s night-time view with the BEA Airtours Comet 4B fleet on the left and the BCAL fleet on the right. (Gatwick Airport Authority)

← By 1973, SAM had replaced its DC-6Bs with former Alitalia Caravelles and retained their basic livery, as seen on I-DABL. (Richard Hamblin)

↓ Alitalia changed its livery in the early '70s, so SAM adapted this when leasing its Caravelles. I-DABT is parked alongside a BEA Airtours Boeing 707-436 in June 1973. (Tom Singfield collection)

↑ A busy, varied line-up in 1979. from front to rear: tail of a Dan-Air Boeing 727, World Airways Douglas DC-8, Britannia Airways Boeing 737, a pair of British Airtours Boeing 707s, and an IAS Cargo Douglas DC-8.

Meanwhile, Dan-Air's DC-3s and Airspeed Ambassadors were nearing the end of their useful life and the airline was looking for a replacement. Dan-Air looked at several types and chose the Nord 262, which could land on short runways and, with one engine running, could achieve a turnround in five minutes using its built-in stairs and had easy access baggage storage. Flight times between the stops on the network could be reduced as it was faster than the DC-3 or Ambassador. It also had a small galley for a drink and snack service to be carried out. However, the introduction of the type did not significantly increase load factors and was not popular, so it was withdrawn after only two years and later replaced with a fleet of Hawker Siddeley HS.748s.

↑ Dan-Air's 1970 timetable featuring the Nord 262A.

← Dan-Air's short-lived sole Nord 262A G-AYFR in January 1971. (John Mounce collection)

➔ By June 1976, there were many BAC One-Elevens operating out of Gatwick including this pair of former British Eagle aircraft, G-ATPJ and ATPJ, now with Dan-Air, seen with a British Airways Viscount.

➔ Much more successful were Dan-Air's Hawker Siddeley HS.748s, including G-AXVG seen in February 1972.

Once British Caledonian had settled down from the merger with BUA and had got the combined fleets repainted in their livery, they were doing well with their One-Elevens and now had four VC10s, these being G-ASIW, G-ASIX, G-ATDJ and also G-ARTA, which had been on lease to Laker Airways until 15 April 1969 when it was acquired by British United Airways just prior to the merger with Caledonian. Readers of a certain age will recall that G-ARTA was the prototype VC10 that wowed spectators at the 1962 Farnborough Air Show wearing BOAC livery. G-ARTA finally adopted the BCAL livery but unfortunately suffered a heavy landing accident in fog at Gatwick on 28 June 1972 during a ferry flight from Heathrow. It was declared uneconomical to repair, so after being hangared and used for spares to keep the other three VC10s airworthy, it was finally scrapped at Gatwick in 1974.

↑ A sad sight as the prototype Vickers VC10 G-ARTA, which last served with British Caledonian, is scrapped at Gatwick in September 1975. (Martin Fenner collection)

Having established his airline on European holiday routes, Freddie Laker now cast his gaze across the Atlantic with plans for an economy service to New York. On 15 June 1971, he applied to the UK Air Licensing Board for the right to launch the world's first low-cost transatlantic service. Fares were to start at £32.50 one way, a third of the cost on regular airlines. Thus began a long legal battle that would not see Laker's Skytrain service begin until 1977.

The request for a licence was initially refused, but granted in February 1972 following an appeal. However, the following month, the UK government revoked this permission and told Laker he needed to apply to the new Civil Aviation Authority instead.

He did so and was granted permission to offer eleven weekly flights between London and New York during the summer, dropping to seven roundtrips in the winter. Laker Airways received its first McDonnell Douglas DC-10s in November 1972, becoming the first airline outside of North America to fly the new widebody aircraft.

However, before flights could begin, the world oil crisis hit in 1973 and airlines began to lose money. British Caledonian faced overcapacity on its North Atlantic routes and warned the government that a new competitor would cause it further difficulty. Hence, Laker Airways' licence to operate transatlantic flights was revoked on 29 July 1975.

Freddie Laker took the government to the High Court, which overturned the decision. Getting US approval caused more delays; however, the airline was eventually granted permission to operate flights for a year on 13 June 1977.

The first Skytrain flight took off from Gatwick for the first time on 26 September 1977, with 272 passengers aboard the airline's 345-seat DC-10. The passengers had paid £59 for a one-way fare on the no-frills flight. There were no advanced reservations, with passengers being sold seats on a first-come first-served, standby basis until the plane was full. By the summer of 1978, the carrier was operating flights with an 80 per cent load factor.

Competitors including Pan Am and TWA were having none of this and, in a move designed to cripple their rival, they matched the airline's standby economy fares. But Skytrain was undaunted: it was a financial success in its first year of operations, encouraging Laker to look at other destinations in the United States.

↑ → Laker Airways advertisements and timetables.

LAKER
Skytrain
TM
No-reservation
AIR PASSENGER SERVICE
A brand-new concept in Transatlantic travel
Laker Airways' Skytrain air passenger service is a brand-new concept in Transatlantic air travel. It is designed to serve travellers forgotten by other air services. It is a *scheduled*, no-reservation, walk-on service, offering at least one flight a day in each direction between London and New York – to a definite timetable, all the year round. It is *the only* service of its kind in the world.
Skytrain air passenger service
LONDON-NEW YORK
£59
NEW YORK-LONDON
$135
(single year round)

WESTERN BELLE

LAKER
Skytrain
Trade mark
AIR PASSENGER SERVICE
Now you can reserve a seat to London on Skytrain!
AIR PASSENGER SERVICE
DAILY SCHEDULED SERVICES
NEW YORK/LONDON/NEW YORK
LOS ANGELES/LONDON/LOS ANGELES
The Laker Airways folder gives you full information on the three tier Skytrain passenger services. The excellent values offered by:
FRAMES HANOVERGRAND
KEMWELL BRITRAIL
COSMOS BIA TOR LINE
will reduce your costs in Britain and Europe.

←↓→ Laker Airways advertisements and timetables.

Passenger Ticket and Baggage Check
833:4200:021
Issued by Laker Airways Limited Gatwick Airport – London, Horley, Surrey.
Skytrain
Trade mark
SCHEDULED AIR SERVICE
LAKER
IMPORTANT: Before travelling you should carefully examine this ticket particularly the Conditions and Notices contained herein.

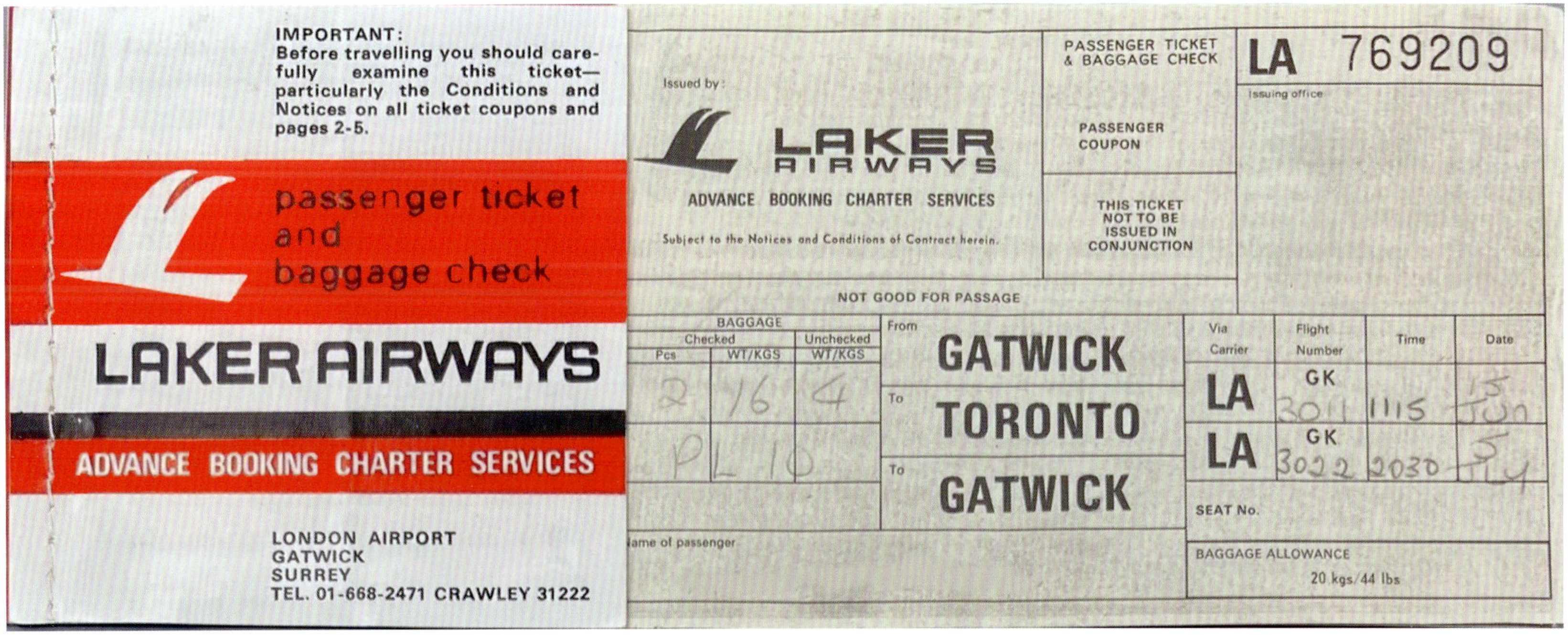
IMPORTANT:
Before travelling you should carefully examine this ticket—particularly the Conditions and Notices on all ticket coupons and pages 2-5.
passenger ticket and baggage check
LAKER AIRWAYS
ADVANCE BOOKING CHARTER SERVICES
LONDON AIRPORT
GATWICK
SURREY
TEL. 01-668-2471 CRAWLEY 31222
Issued by:
LAKER AIRWAYS
ADVANCE BOOKING CHARTER SERVICES
Subject to the Notices and Conditions of Contract herein.
PASSENGER TICKET & BAGGAGE CHECK
LA 769209
Issuing office
PASSENGER COUPON
THIS TICKET NOT TO BE ISSUED IN CONJUNCTION
NOT GOOD FOR PASSAGE
BAGGAGE
Checked
Unchecked
Pcs
WT/KGS
From
GATWICK
To
TORONTO
To
GATWICK
Via Carrier
Flight Number
Time
Date
LA
GK
LA
GK
SEAT No.
ame of passenger
BAGGAGE ALLOWANCE
20 kgs/44 lbs

To accommodate such long-haul flights, in 1973 Gatwick's runway had been extended again to 3,098m, enabling non-stop flights to the US West Coast.

In 1973, Dan-Air received its first Boeing 727. Specifically, the first of three former Japan Airlines 727-100s arrived on 13 April 1973, making Dan-Air the first British airline to operate the type. As the 727 fleet was increased, some of the early Comets were withdrawn from service, including Comet 4B G-APDM, which had its wings removed in the early '70s and continued to serve as a ground instructional cabin trainer.

↓ Passengers boarding Dan-Air's first Boeing 727-46, G-BAEF, in May 1973.

→ Dan-Air Boeing 727 passenger flight guide brochure c.1973.

↑ With wings chopped off, Dan-Air's Comet 4 G-APDM is seen here in use as a ground cabin trainer on 26 July 1978. (W.F. Wilson)

Smaller European domestic airlines were still much in evidence at Gatwick in the 1970s, operating older turboprop and jet aircraft types rarely seen at other UK airports. This indeed was a 'golden age' for spectators and plane spotters, with the viewing terraces being very popular.

→ The spectators' viewing terraces and piers were still very popular, as seen in this 1974 view. (Courtesy Tony Lewis)

↑ Fokker F.27-200 Friendship LN-SUE of the Norwegian airline Braathens SAFE arrives on 25 July 1971.

↑ When British Airways was formed from the merger of BEA and BOAC on 1 April 1974, BEA Airtours became British Airtours and painted its Boeing 707-436s in the 'Negus' livery, as seen on G-APFJ.

A busy scene in 1970 as passengers disembark BEA Vickers V.802 Viscount G-AOJB and cross the apron.

On 18 March 1978, Braniff International launched a daily non-stop Boeing 747 flight from Gatwick to Dallas/Fort Worth. This was the airline's first nonstop service from the US Southwest to the European Continent. The inaugural flight was flown by Boeing 747-127 N601BN, which was called 'the Big Orange' due to its bright livery and was quite an attraction.

↑ 'Big Orange Comes to Town!' This caused quite a stir when Braniff International started operating a daily non-stop service into Gatwick from Dallas/Fort Worth with Boeing 747-127 N601BN, as seen on 10 September 1978. (W.F. Wilson)

This service was followed by a series of US airlines operating into Gatwick, including Delta Airlines, who, on 1 May 1978, launched a daily flight from Gatwick to New Orleans via Atlanta, which was operated by Lockheed L-1011 TriStars.

↓ → Delta Air Lines began its service to Gatwick in June 1979. They initially operated the route with leased Lockheed L-1011-100 TriStars before introducing their own longer-range -500 series aircraft later that year, including N756DR.

In the mid-70s, the terminal building was extended again, the new extension being officially opened by Prince Charles on 9 June 1978. He arrived by train, this cementing a long and profitable relationship between the railway and the airport. Also on that day, a helicopter service with two Sikorsky S-61Ns was established linking Gatwick with Heathrow for connecting passengers.

← ↑ → Before the M25 was completed, the Airlink helicopter service transported passengers between Gatwick and Heathrow airports using two Sikorsky S-61Ns.

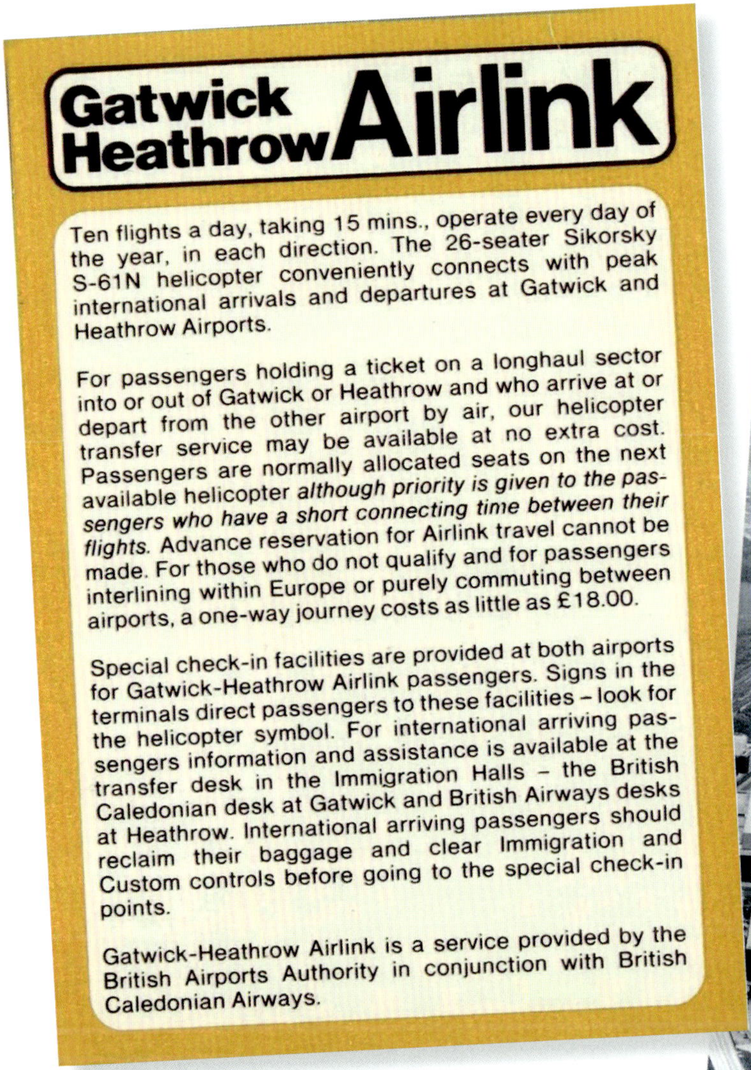

Gatwick Heathrow Airlink

Ten flights a day, taking 15 mins., operate every day of the year, in each direction. The 26-seater Sikorsky S-61N helicopter conveniently connects with peak international arrivals and departures at Gatwick and Heathrow Airports.

For passengers holding a ticket on a longhaul sector into or out of Gatwick or Heathrow and who arrive at or depart from the other airport by air, our helicopter transfer service may be available at no extra cost. Passengers are normally allocated seats on the next available helicopter *although priority is given to the passengers who have a short connecting time between their flights.* Advance reservation for Airlink travel cannot be made. For those who do not qualify and for passengers interlining within Europe or purely commuting between airports, a one-way journey costs as little as £18.00.

Special check-in facilities are provided at both airports for Gatwick-Heathrow Airlink passengers. Signs in the terminals direct passengers to these facilities – look for the helicopter symbol. For international arriving passengers information and assistance is available at the transfer desk in the Immigration Halls – the British Caledonian desk at Gatwick and British Airways desks at Heathrow. International arriving passengers should reclaim their baggage and clear Immigration and Custom controls before going to the special check-in points.

Gatwick-Heathrow Airlink is a service provided by the British Airports Authority in conjunction with British Caledonian Airways.

➔ An aerial view of Gatwick in the 1970s. (Gatwick Airport Authority)

Piston-engined airliners seen at Gatwick were becoming rarer by the mid-'70s and one of the last operating passenger flights was DC-6A HB-IBS of Swiss airline Balair, which was the last DC-6 variant in airline passenger service in Europe. It was constructed as a DC-6A (with a main-deck freight door) being delivered to the Northern Aircraft Operating Company as G-APNO in June 1958. Following service with Hunting Clan Air Transport, it later joined British United Airways, following the airline's formation. Balair acquired it in January 1969, registered HB-IBS, and subsequently it was used solely for passenger flights until being withdrawn and sold in 1982.

In May 1977 the late propliner enthusiast Stephen Piercy proposed chartering the aircraft for a one hour enthusiast's flight for £15, while another offered a one-hour flight in a different DC-6 from Gatwick for £9.95. The Balair flight went ahead on 21 May 1977 and lasted forty-seven minutes, taking off from Gatwick an flying via Lydd and return. Clearly there were a lot of takers!

↑ One of the last chances to sample a flight in a large piston-engined airliner from Gatwick was snapped up by these happy aviation enthusiasts on Balair's Douglas DC-6B in May 1977. (Tom Singfield).

↑ One of the last chances to sample a flight in a large piston-engined airliner from Gatwick was snapped up by these happy aviation enthusiasts on Balair's Douglas DC-6B in May 1977. (Tom Singfield).

In 1979, Laker Airways began to deploy McDonnell Douglas DC-10-30s on a new route between Gatwick and Los Angeles. Following on from this, they introduced winter flights from Gatwick to Miami via Bangor.

In September 1978, Laker became the UK launch customer for the Airbus A300 when they ordered ten of the type for a short-lived scheduled Skytrain service between Manchester and Zurich, also aiming to offer low-cost flights between the UK and other European destinations.

↑ Laker Airways Airbus_A300B4-203, G-BIMB in August 1982. (Martin Fenner collection)

Handley Page Heralds were present in prolific numbers at Gatwick over the years, starting in 1962 with Jersey Airlines, which merged into the British United Airlines group as British United (CI) Airlines (Channel Islands) and then changed its name to British United Island Airways in 1967. In 1970, the airline became British Island Airways and operated a large fleet of Heralds, followed by BAC One-Elevens. In 1991 it merged with Air Anglia, Air Wales and Air Westward to form Air UK, at the time the UK's biggest regional airline and its third-largest scheduled operator.

↑ The Handley Page Herald was a prolific type at Gatwick throughout its life. Here are G-APWE and G-ASKK of British Island Airways parked together on 19 April 1976. BIA operated up to twenty Heralds until 1982, when it merged with Air Anglia to form Air UK. (Richard Hamblin collection)

FIVE

1980S: THE AIRPORT EXPANDS WITH NEW COMPETITION AND SOME FAREWELLS

Unfortunately, Laker Airways collapsed after the early recession of the 1980s and Laker only received three Airbus A300B4 aircraft of the larger order of ten aircraft that were intended to be used for the Skytrain service to Europe. The airline took delivery of the first A300B4, G-BIMA, in January 1981, which operated its last flight back on 5 February 1982; the same day it went bankrupt. British Airways, British Caledonian, Pan American, Trans World Airlines and Air Florida agreed to honour tickets for approximately 4,000 displaced passengers.

Another airline that was to become a major force at Gatwick was Air Europe, which formed in 1979 as a small charter company with three brand-new Boeing 737-200s. It was the main supplier of charter seats to Intasun Leisure, the UK's second-largest package tour operator after Thomson.

Air Europe would go from strength to strength in the 1980s, purchasing more Boeing 737s and becoming a major presence on short-haul European trunk routes that served Gatwick. It acquired more landing slots, which enabled it to increase both the frequency of routes and the amount of countries it served. By the end of the decade it was Gatwick's largest resident airline with a large fleet of Boeing 737s and 757s.

Alas, this success was not to last. The cost of borrowing to purchase more aircraft rose and its parent company would be beset by financial difficulties by the end of the decade. Air Europe went bankrupt on 8 March 1991.

↓ A line-up of Air Europe Boeing 737-300s including G-BNPB and LN-NOS on 29 May 1989.

On 1 August 1980, British Caledonian Airlines took on the British Airways monopoly route to Hong Kong, flying daily from Gatwick with their DC-10s. They took their advertising seriously at both ends of the route by even having two of Hong Kong's trams painted up in their colours.

↓ → On 1 August 1980, BCAL launched a DC-10 service from Gatwick to Hong Kong, competing with British Airways, who previously had a monopoly on this route. BCAL's advertising extended to having two of Hong Kong's trams painted in their livery.

British Caledonian
Passenger Ticket and Baggage Check issued by
BRITISH CALEDONIAN AIRWAYS LIMITED
N 227855
PASSENGER COUPON
Let's go
BRITISH CALEDONIAN
RITISH CALEDONIAN DC10-30
Hong Kong
British Caledonian
British Caledonian
Boarding
BALCH M
23A
British Caledonian
Boarding Pass

144
British Caledonian. A better way to London.
英國金獅航空公司
British Caledonian Airways
We never forget you have a choice
Silver

Gatwick-based British Island Airways (BIA) became a part of Air UK in 1980 as a result of a merger of four UK-based regional airlines, including Norwich-based Air Anglia. Operations began on 16 January 1980, at which time Air UK was the largest regional airline in the UK and the country's third-largest scheduled carrier, with a staff of 1,700 and over 1 million passengers a year. It had a fleet of forty aircraft, comprising two F-28 4000 series Fellowships and ten Fokker F27 100/200 series Friendships from Air Anglia; four BAC One-Eleven 400s and eighteen Handley Page Dart Heralds from BIA; and six Embraer 110 Bandeirantes that originally part of the BIA, Air Wales and Air Westward fleets. Apart from the four One-Eleven 400s, which were predominantly operated on charter flights, all the other aircraft were part of Air UK's scheduled service fleet.

↑ Air UK Embraer EMB-110 Bandeirante G-BKBG taxies past Laker's fleet of withdrawn DC-10s and a Boeing 707 in July 1982.

↓ British Island Airways BAC One-Eleven 406AY G-AXBB was leased to Air Florida as a connecting service for their DC-10 schedule in June 1984.

In 1981, public enquiries started into the requirement for a second terminal building, resulting in planning consent being granted for construction of the North Terminal by the end of 1982.

Meanwhile, a satellite pier was added to the terminal building, connected by the UK's first automated people mover. Construction of the North Terminal began in April 1983, as well as work on a new control tower, fire station and an on-site Hilton Hotel. There was still a requirement for a second runway, but due to pressure from local communities the Airport Authority agreed to convert taxiway 1 into an emergency runway for use only should the main runway be closed for any reason. This work was carried out in 1984 and was used during the winter of 1985–86, while the main runway underwent some long-term maintenance and resurfacing.

The air traffic control tower in the original Beehive terminal closed in the late 1950s and a new control tower was built in the middle of the airfield that has been in constant use ever since. It was upgraded with extensions and additions over the years. In 1984, a new replacement air traffic control tower was opened, which is equipped with advanced technology and systems, such as radar, radio, meteorological instruments, and flight data processing. The tower is on the south side of the airport, near the main terminal building, and has a height of 37m, which was the tallest in the UK at the time. It offers a panoramic view of the airport and its surroundings and is current today.

↓ On 7 November 1981, BCAL BAC One-Eleven G-ASJI is parked in front of a circular satellite pier under construction, which was connected to the main terminal by the UK's first automated people mover. (Stephen Wolf)

←↓→ By the early '80s, Gatwick's air traffic control operation looked like this. Here controllers are at work in the visual control tower and radar room downstairs. (Gatwick Airport Authority)

→ Gatwick's new Air Traffic Control tower was opened in 1984. (NATS photo)

↓ A mid-'80s aerial view of Gatwick. (Mike McBey, Gatwick Airport)

After Freddie Laker's low-cost Skytrain failed, Sir Richard Branson launched Virgin Atlantic Airlines to have a go at competing with British Airways and American Airlines on the transatlantic sector. The airline's initial launch, with a leased Boeing 747, marked the beginning of a unique and innovative airline that challenged the status quo. The airline's first flight was on 22 June 1984, from Gatwick to New York Newark.

→ Richard Branson launches Virgin Atlantic Airlines in 1984.

↓ Boeing 747 G-VIRG was Virgin Atlantic's first aircraft. It was a 747-287B and was affectionately nicknamed 'Maiden Voyager' by Richard Branson. This aircraft was instrumental in the early years of Virgin Atlantic, marking the start of a unique naming tradition. (W.F. Wilson, 19 August 1984)

Following redevelopment and expansion of the South Terminal, the excellent viewing piers were closed on security and safety grounds, and in the mid-1980s plane spotters and spectators were moved on to the roof terrace of the re-opened South Terminal. This was quite understandable and acceptable with a café and souvenir shop established there. Things looked good and clearly the British Airports Authority still respected and appreciated the public relations value of the airport through the visiting public.

In February 1986, the M25 motorway link between Heathrow and Gatwick was completed, resulting in the withdrawal of the helicopter Airlink service with its two Sikorsky S-61Ns.

Construction of the North Terminal began in 1983 and it was opened in 1988. In the late '80s, British Airways took over British Caledonian, the process beginning on 21 December 1987 and being completed on 14 April 1988. On 18 March 1988, the North Terminal and the automated rapid transit system link to the South Terminal were officially opened by Queen Elizabeth II.

↓ ↑ The Queen officially opening the new North Terminal on 18 March 1988. (Gatwick Airport Authority).

← ↓ The automatic Shuttle Transit commenced running between the North and South Terminals in 1987. (Gatwick Airport Authority)

By now, Gatwick had become the world's second busiest airport in terms of international passengers. Dan-Air was carrying millions of passengers on scheduled and charter services across the world. It achieved several firsts including introducing the first Boeing 727 to the UK register, being the world launch airline of the BAE 146 'Whisperjet' and operating the world's largest fleet of the legendary Comet airliner.

↓ → In March 1983, Dan-Air became the launch customer for the British Aerospace 146. G-BKMN is about to line up to depart on 28 April 1984. (W.F. Wilson)

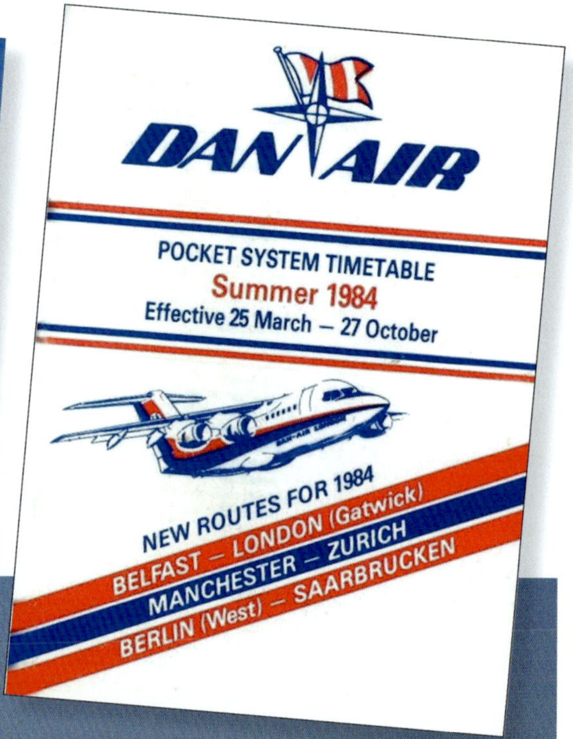

In 1983, British Caledonian started a charter subsidiary called BCAL Charter with some of their DC-10s. In 1985, it was rebranded as Cal-Air, then in 1988 changed to Novair. This only lasted for another three years as Novair ceased operations in 1990.

→ A spectacular 1981 line-up of six BCAL DC-10s, of which two have British Caledonian Charter titles, and a Wardair DC-10 at the end of the line.

↓ Cal-Air was formed as a charter airline by British Caledonian using two former Laker Airways DC-10-10s in 1982. G-BJZD is seen about to land on Runway 26 on 17 April 1987.

↑ Two years later, a near identical photo captured G-BJZD on 29 May 1989 in a similar livery but now with Novair International Airways, which formed on 7 December 1988 following the sole acquisition of Cal Air International by the Rank Organisation.

The 1980s saw more American airlines start operating into Gatwick. In 1980, Northwest Orient launched services to Minneapolis St Paul with Boeing 747s and DC-10s.

Western Airlines began serving Gatwick from Anchorage and Denver in January 1981, initially using a single DC-10-30 aircraft. Later that same year, they also added a non-stop London to Denver service with a second DC-10-30.

This was followed on 4 April 1981, when Air Florida commenced a new service to Miami operated with a DC-10 aircraft. Trans World Airlines then launched non-stop flights to New York on 26 April 1982 using Lockheed TriStars, a service they were already offering from Heathrow. They also introduced a daily non-stop flight from Gatwick to St Louis on 29 April 1985.

Not to be outdone, American Airlines commenced services in May 1982 with Boeing 747s to Dallas/Fort Worth and People Express also started operating Boeing 747s to New York Newark.

↗ Another glorious Gatwick ramp line-up on 9 November 1980 includes British Airtours Boeing 707-436 and Northwest Orient Boeing 747-151 N601US. (Stephen Wolf)

→ Western Airlines flew a single DC-10-30, N821L named 'The Londoner', on a short-lived Denver–London service in 1981. It is seen alongside BCAL DC-10 G-BEBL and Laker DC-10 G-AZZC, all dwarfing a Britannia Airways Boeing 737, on 7 November 1981. (Stephen Wolf)

Continental Airlines first began serving Gatwick on 28 April 1985 with a non-stop flight from Houston, which was the airline's first scheduled service to Europe. The airline later expanded its service with flights from Newark and other US cities, particularly after merging with People Express in 1987. Continental continued to operate from Gatwick until its merger with United Airlines in 2012, when all Continental flights were transferred to United.

➔ Former People Express Boeing 747-143 commenced operations with Continental Airlines in to Gatwick in 1985. This photo was taken on 8 April 1990, but it is still wearing the early gold and red livery.

➔ Trans World Airlines served London Heathrow for years, but started a Seattle to Gatwick service in 1984. Here is Boeing 747-131, departing on 18 August 1993.

Airlines also began to serve the Far East from Gatwick. Cathay Pacific Airways launched three times weekly Boeing 747 flights from Gatwick to Hong Kong via Bahrain on 17 July 1980, increasing the frequency to daily a year later. Philippine Airlines (PAL) also started services in 1980, as part of a broader expansion that included flights to other European cities like Amsterdam, Athens, Frankfurt, and Rome. However, PAL later discontinued its European routes, including Gatwick, before eventually returning to London with a service to Heathrow in 2013.

→ In 1980, Cathay Pacific operated its first flight between Hong Kong and Gatwick, with a fuel stop at Bahrain. The route was initially flown thrice weekly, but increased to a daily schedule from 1 July 1981. Boeing 747-367 VR-HII is on push-back about to depart for Hong Kong on 9 March 1986. (The late Roger Braithwaite)

↓ Philippine Airlines was another Boeing 747 operator that commenced services into Gatwick in 1980. Boeing 747-211B N207AE is arriving from Manila on 4 May 1993.

Farewell to the Comet

Dan-Air was so successful for so long mainly because it didn't rush to get financial backing to buy the latest jet airliners. Instead, it carefully sourced reliable second-hand aircraft available for sale as soon as airlines put them up for disposal. These were mainly of British design such as the BAC One-Eleven, Hawker Siddeley HS.748, and Vickers Viscount, types supported by the British aircraft and engine industry at the time, but it was the de Havilland Comet that the airliner might be most remembered for.

Despite early setbacks with accidents involving the early Comet 1, the type was developed into a safe, reliable icon of the British aircraft industry. It was not only the first British commercial jet airliner but the first jet airliner to cross the Atlantic, when BOAC beat Pan Am with their Comet 4 in 1959. There were 114 Comets built of all variants, including 74 4/4Cs and 18 4Bs. They were delivered to airlines around the world – Aerolineas Argentinas, East African Airways, Kuwait Airways, Malaysia-Singapore Airlines, Sudan Airways, MEA of Lebanon, United Arab Airlines (Egypt), Mexicana, Olympic Airways and, of course, BOAC and BEA.

When BOAC retired its Comet 4s, Dan-Air was quick to acquire them in 1966. It liked the type so much that when other airlines around the world withdrew theirs from service, it would quickly purchase those, which would be flown into nearby Lasham and completely refurbished and turned out in the airline's livery. Dan-Air acquired a total of forty-nine Comets between 1966 until 1980, with some bought just for spares. When the RAF retired its five Comet 4Cs (known as C.Mk.4s) in 1975, Dan-Air were delighted to purchase them, as they had low airframe hours, being used as military transports by 216 Squadron at RAF Lyneham since 1962.

↑ Dan-Air Vickers V.807B Viscount G-CSZB sits awaiting its next service in April 1980.

↑ By August 1980, Dan-Air's Comet 4B G-APMB had been withdrawn from service, had its wings removed and was used as a ground handling trainer for several years. (Stephen Wolf)

One former BEA Comet 4B was grounded and used as a cabin crew trainer, while another lingered on beyond 1980 with its wings removed and was used for ground handling training. Gatwick was the last place to see operational Comets until the last one, G-BDIW (formerly XR398 in RAF service), was retired on 9 November 1980. It later made a special farewell flight from Gatwick and was then donated to a museum in Hermeskeil, Germany.

In 1986, Dan-Air entered the wide-body era with a pair of 336-seat Airbus A300s purchased from German airline Hapag-Lloyd, which were used on high-density routes to the Canaries and Balearics in Spain and to Portugal. This gave Dan-Air the opportunity to compete with rival airlines who were already operating wide-body aircraft. However, a downturn in charter flights faced by all airlines in the late 1980s led to Dan-Air selling its A300 four years later.

↑ A sad day in November 1980 when Dan-Air retired its last Comet after fourteen years of Comet operations. Comet 4C G-BDIW was subsequently flown to Germany for preservation by the Hermeskeil Museum. (Tom Singfield)

↓ Dan-Air's last type in service was the Airbus A300. This A300B4-203, G-BMNB, was purchased from Hapag-Lloyd and is seen in April 1987.

After nearly twenty years at Gatwick, some were quite shocked when, in 1988, British Airways took over British Caledonian. The resulting conditions imposed by Britain's Monopolies and Mergers Commission following the takeover meant that BCal's former role as Gatwick's prime scheduled short-haul operator was taken over by Dan-Air and Air Europe. Meanwhile, British Airways took up BCal's previous role as the airport's main long-haul carrier. In the early 1990s, British Airways purchased Dan-Air and on 8 March 1991 Air Europe ceased operations, enabling British Airways to build up Gatwick as a secondary hub to London Heathrow and become the dominant airline by 2000.

The merger included British Airtours changing its name to Caledonian Airways and using the TriStars that had formerly belonged to BA. BA's subsequent decision to de-hub Gatwick provided the space for easyJet to establish its biggest base at the airport and become the dominant airline.

In 1989, plane enthusiasts heard that a rare flying boat was coming to Gatwick! They wondered if the airport was going to flood the runway, but when the world's last airworthy Short Sunderland, G-BJHS owned by Edward Hulton, arrived on 31 July 1989, it was flown down the runway in spectacular style by Ken Emmott. Wearing Ryanair titles, was this going to be a new service from this budget airline that would surely include an extra baggage allowance? Alas no, just a short sponsorship deal!

Following this especially spirited flypast, Emmott received a mock bill from Gatwick: £20,000 'To repair runways, namely filling in the grooves left by low flying Sunderland', £1,500 'For treatment to shock to unspecified number of spectators', and lastly £50, 'Laundry bill for above …'! This aircraft was later sold and is now in Kermit Weeks' Fantasy of Flight Museum in Kissimmee, Florida.

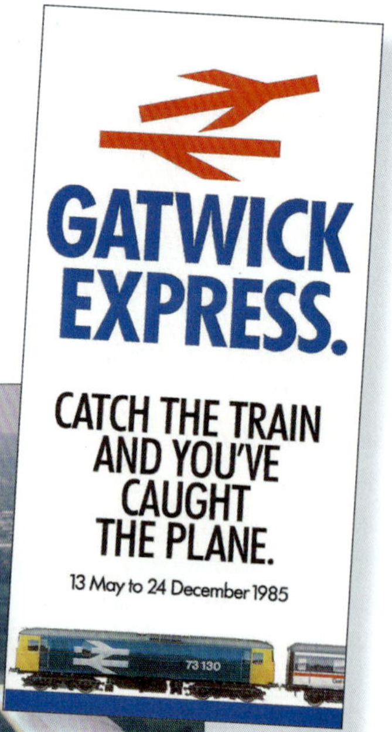

→ British Rail Gatwick advert.

↓ The British Caledonian Airlines offices and hangar with the original Beehive terminal visible in the distance, photographed from a Dan-Air BAC One-Eleven in June 1984. (Vic Cook)

← 'Stand by and we'll flood the runway!' A magical shot of Edward Hulton's Short Sunderland G-BJHS making a flypast at Gatwick on 31 July 1989. (Alan Tirrell, Gatwick Airport official)

SIX

1990S: COMPETITION GETS SERIOUS

Throughout the 1990s, Gatwick Airport saw significant developments, including the opening of the North Terminal's second pier and the expansion of both the North and South Terminal international departure lounges. Additionally, the airport's main runway was extended, and easyJet began operations from Gatwick.

On 1 July 1991, Virgin Atlantic began operating from Heathrow. This marked a significant milestone for the airline, which had previously been limited to Gatwick for its first seven years of operation. The decision to allow Virgin Atlantic into Heathrow was influenced by the abolition of the London Air Traffic Distribution Rules (TDRs), which had favoured Gatwick.

Now Virgin Atlantic had the rights to operate from both Gatwick and Heathrow, despite objections from British Airways and American Airlines, who saw this competition impacting on their traffic. Richard Branson strongly defended this and even marked his aircraft with 'No Way BA/AA' in defiance, with a judge ruling in his favour in that competition was healthy to keep prices down. British Airways tried everything to thwart Virgin, including corporate espionage and libel, but Branson took them to court in the high-profile 'Dirty Tricks' lawsuit and in 1993 won damages worth £610,000, which he distributed to his staff. Consequently, Virgin Atlantic launched several new routes, including daily flights to New York JFK, Los Angeles and Boston.

Today, Virgin Atlantic is the third largest UK airline and currently operates out of both Gatwick and Heathrow.

↑ Virgin Atlantic's Boeing 747-287Bs G-VIRG and G-VLAX on 17 August 1998 wearing 'No Way BA/AA' logos, objecting to British Airways' and American Airlines' monopoly on the UK–USA routes.

After nearly forty years of operations, it was a sad day at Gatwick in 1992 when the name Dan-Air was no more, with the airline being taken over by British Airways.

Regional airlines should not be overlooked, especially those operating turboprop airliners in the 1990s. One of those was CityFlyer Express, which was a short-haul regional airline, based at Gatwick operating ATR-42s and later ATR-72s.

In 1993, operating at British Airways Express, CityFlyer became the first British Airways (BA) franchisee. In 1999, CityFlyer's ownership was acquired by BA, when that company purchased the original promoters as well as 3i, which was the airline's main shareholder at the time. Initially, CityFlyer continued to operating as a separate airline, but in 2001, it was finally absorbed into British Airways' mainline short haul operation at Gatwick, which resulted in a change in British Airways' strategy for its Gatwick operation. Following its integration into British Airways, the airline's ATRs and Short 360 turboprops were withdrawn.

↗ CityFlyer Express became a franchise of British Airways in August 1993, the first franchise agreement in the UK airline industry. This initial agreement was for five years and saw CityFlyer operate under the 'British Airways Express' banner, with their aircraft being repainted in the smart BA 'Landor' livery. ATR-42 G-BUEA is about to land on runway 08 on 9 April 1995.

→ Although Monarch Airlines was based at Luton Airport, it established a second base at Gatwick in 1981 and ordered four Airbus A300B4s to specifically operate from the airport on holiday routes. These were delivered in 1990–91, one of which, G-OJMR, is seen alongside an Air Malta Boeing 737 on 18 August 1993.

A Surprise Visitor!

On 2 April 1996, Gatwick hosted a surprise visitor in the shape of Air France Concorde F-BTSD. If that was not a surprise in itself, it was in full Pepsi-Cola livery.

Faced with a falling market share, Pepsi-Cola decided on a $500 million rebranding and started to look around for a dramatic way to advertise its new branding. It decided on an advertisement involving Concorde.

Pepsi put proposals to both Air France and British Airways, the only two operators of the supersonic airliner, resulting in the contract being awarded to Air France. As such a dark colour tends to build, retain and hold heat longer than light ones, the issue of temperature was very important to Concorde. Consequently, Air France maintenance staff needed to contact Aerospatiale, the French joint manufactures of Concorde, as the aircraft is only certified to be painted overall in a white livery. Approval was received to paint the fuselage in blue, but they were advised to keep the wings white due to fuel temperature concerns. In March 1996, painting the aircraft commenced at the Air France maintenance facility of Paris, Orly, requiring 200 litres of paint and 2,000 hours of work to complete the tasks.

As this Concorde was due to be presented in British Airways backyard, so to speak, Air France required its name to appear near the cockpit, as well as the seahorse logo, this in addition to the Pepsi blue paint scheme.

The whole operation carried out to this Concorde was to be undertaken in secret, as Pepsi wanted to keep it as a surprise for the moment when it would unveil its new identity, Pepsi Blue. Concorde 'Sierra Delta' was thus covered by brown wrapping paper after it was painted, which would allow as few people as possible to be aware of the new paint scheme. It eventually left the hanger on 31 March during the night, and was quickly rolled on to the runway. It then took off for Gatwick, where Pepsi was to receive its guests at a special launch event. On arrival, the aircraft was immediately towed into a hangar for the night.

The unveiling took place on 2 April 1996, with star guests including Claudia Schiffer, Andre Agassi, Cindy Crawford and hundreds of journalists invited by Pepsi to the event. After the event, Concorde departed on a promotional tour visiting major airports across Europe and the Middle East, arriving back at Paris Orly on 9 April.

↑ Surprise visitor: Air France Concorde F-BTSD in the short-lived Pepsi-Cola livery on 31 March 1996. (Mike Axe)

↑ The Pepsi-liveried Concorde departing Gatwick on 2 April for a short European promotional tour.

Founded by Stelios Haji-Ioannou to offer low-cost fares in Europe, easyJet made its first flights in November 1995 flying from Luton to Glasgow and Edinburgh.

It began operating from Gatwick in 1999, flying to Geneva on 16 December. Within a month, Edinburgh, Zurich, Malaga and Palma were all added to the LGW route map. On 1 February 2002, easyJet opened a base at Gatwick, with its first based flight being a Boeing 737 service to Barcelona. To mark the opening, the airline gave Boeing 737 G-EZJD a 'Big@Gatwick' logo.

↓ Another Luton-based airline, easyJet, began operating from Gatwick on 16 December 1999, later establishing a base there on 1 February 2002. To mark this milestone, Boeing 737-73V G-EZJD carried the logo 'BIG@Gatwick' on its port side.

Since then, for more than two decades, easyJet has flown more than 200 million passengers to and from the airport and expanded, becoming the airport's largest airline client. Despite having been founded with its corporate headquarters at London/Luton, Gatwick is now officially the largest base anywhere in the easyJet network.

British Airways tried to compete by creating a low-cost subsidiary airline in 1998 called Go Fly. Try as they might, they could not compete and easyJet subsequently absorbed the Go Fly fleet of Boeing 737s. It went from strength to strength adding more 737s, followed by Airbus A319s, A320s and A321s.

On the evening of 31 July 1994, Channel Express Handley Page Herald G-CEXP was about to take off from Gatwick on a mail flight to East Midlands Airport. As the pilot opened the throttles, he noticed that the needles of temperature gauges for the engines (two Rolls-Royce Dart turboprops) were against the stops and in excess of 1,000°C. Thus he aborted the take-off and moved the Herald from the runway onto the adjacent taxiway. As he was shutting down the engines the port engine fire warning light illuminated and so he activated the engine extinguisher. The crew vacated the Herald as the airport fire engines arrived. On subsequent examination, both engines were found to be seized, with metal debris in their jet pipes.

↓ Handley Page Herald 401 G-CEXP of Channel Express seen in happier times at Hurn on 2 April 1994. (Stephen Wolf)

This particular Herald first flew in 1968, the aircraft being delivered to Israeli carrier Arkia as 4X-AHO. It later served as G-BFRJ with Express Air Freight, I-ZERC with Soc. Columbia and as G-CEXP with Channel Express from 1987.

Even in 1994, the Herald was a forty-year-old design, so, it came as a surprise to no one that the incident in July 1994 marked the end of the road for the aircraft. However, rather than being reduced to parts or being scrapped, it became a visitor attraction on the roof of the South Terminal at Gatwick where there still was, in less security conscious times, viewing terraces.

After refurbishment, the Herald was hoisted onto the terrace in 1996 and was opened up to the public. This was accompanied by the nose of a de Havilland Comet 2, which was built for BOAC as G-AMXA. It was then sold to the Ministry of Supply in March 1955, one of four converted to C2R standard that joined the RAF in the ELINT role as XK655 with 51 Sqn at Wyton, where it operated from 1963 until its retirement in 1974. Flown to Strathallan in Scotland in August 1974 for preservation, it was finally broken up in 1990 after the collection was dispersed. The nose was saved, painted back into BOAC colours, and put on display on the Sky View Terrace, with its cockpit able to be viewed from the rear.

In 1998, the main runway was further extended to 3,316m, followed by both terminals being extended at a cost of £60 million.

↓ Herald G-CEXP on 30 March 2002 after being mounted on top of the South Terminal viewing terrace on 26 March 1996. It was donated by Channel Express and converted into a walk-through exhibit for the Skyview display. It remained there until the viewing terrace closed on 10 September 2003.

↑ Alongside the Herald was mounted the nose section of Comet 1A G-AMXA in BOAC colours. The cockpit could be viewed from the rear.

↑ A cluster of Northwest DC-10s including N225NW await passengers for their next departures to the US on 17 August 1998.

SEVEN

2000S: NEW ROUTES, NEW AIRCRAFT AND FURTHER EXPANSION, BUT SPECTATORS NO LONGER WELCOME!

In 2000, both terminals were extended at a cost of £60 million, adding seating, retail space, and catering outlets. This was followed by the North Terminal's departure lounge being extended in 2001.

In June 2003, Pier 6, which is an eleven-pier stand, was attached to the North Terminal by a remarkable bridge. The development required an investment of £110 million and was constructed off site and moved into position in May 2004. The entire Pier 6 development, including aircraft piers and retail concessions, opened a year later.

On 29 December 2000, a male passenger entered the flight deck of British Airways Boeing 747 flight BA2069 en route from Gatwick to Nairobi and attempted to seize the flight controls. The flight crew and captain struggled with the passenger, during which the auto-pilot became disengaged and the plane dropped a few thousand feet. Fortunately, the crew managed to both restrain the man and land the aircraft safely.

The events of 11 September 2001, when terrorists hijacked two American Airlines and two United Airlines flights in the United States, have been well documented. Two of the aircraft were crashed into the World Trade Center's twin towers in New York. A third aircraft crashed into the Pentagon, while the fourth crashed into woodland in Pennsylvania.

Following these incidents both BA and Virgin Atlantic took radical action, including the grounding of aircraft and suspension of routes. This also had a major impact on security at airports and on board airliners. At Gatwick, the roof terraces were closed and no longer could plane spotters, or indeed any potential terrorists, get so close to the aircraft.

By 2003, the 'golden age' for spectators and plane spotters at Gatwick Airport was over, when the Skyview Terrace was closed and the Herald and Comet nose were removed, which had been an attraction since 1996. The Comet nose found a new home when it was transferred to the Al Mahatta Museum in Sharjah, where it survives today, but the Herald was dumped unceremoniously at the side of the airfield for the fire service's use. However, it lingered on there for years and could be seen by passengers in all arriving and departing aircraft. Several attempts were made to rescue it and move it to the nearby Gatwick Aviation Museum, but all failed on the grounds of cost, so as this is written its future remains uncertain.

↓ A sad sight on 1 July 2011 with the former Skyview terrace's Herald G-CEXP dumped ignominiously at the side of the airport. Plans to preserve it had been abandoned at the time of writing.

Airlines come and go at Gatwick; some run into financial difficulties, while others are subject to mergers or takeovers by other airlines, or just cease operations for other reasons. It is just not possible to include them all within the scope of this book.

One example was Excel Airways, which later became XL Airways UK. This was a British low-cost charter and scheduled airline that operated from 1994 to 2008, when it ceased operations due to financial difficulties within its parent company, XL Leisure Group. The airline was known for offering affordable package holidays to sunny destinations.

↓ A classic line-up on 16 May 2006 with Excel Airways Boeing 757-250 G-VKNA alongside a another 757 of Continental Airlines and a Boeing 767 of Thomson Airways.

In 2009 the British Airports Authority was forced by the Competition Commission to sell Gatwick to Global Infrastructure Partners (GIP) for £1.5 billion, which announced a £2.5 billion investment programme.

In May 2016, easyJet opened its self-service bag drop area at Gatwick, which includes automated check-in kiosks and bag drop machines. This area is the world's largest, boasting forty-eight automated bag drop machines within a 5,000sq m space. The opening marked a major milestone in easyJet's move to a single-terminal operation at Gatwick.

➔ easyJet's self-service check-in area in April 2025.

↑ easyJet Airbus A320s in their current livery await their passengers on 13 June 2019.

↑ In February 2022, easyJet celebrated twenty years at Gatwick. (easyJet photo)

↑ Today Gatwick is dominated by TUI and easyJet, as seen in this aerial line-up on 3 June 2021. (Ian Haskell)

→ An easyJet passenger's view of Airbus A320s in their previous livery on 30 May 2010.

On 6 July 2012 the world's largest airliner arrived, when an Emirates Airbus A380 flew the type's first scheduled service into Gatwick as a one-off special event to mark the airline's twenty-fifth anniversary at the airport, and to test the aircraft's suitability for the airport. To cater for regular A380 services, two compatible stands were completed in late February 2013, enabling jet bridge access from the west end of the North Terminal's Pier 6.

Norwegian Air Shuttle established its first UK base at Gatwick in May 2013, initially basing three Boeing 737-800 aircraft. During the first quarter of 2014, Norwegian had a market share of 10 per cent at Gatwick.

↓ Norwegian Air established its first UK base at Gatwick in May 2013. While they began operating some routes to Gatwick earlier, this marked the official establishment of their base. By the first quarter of 2014, they held a 10 per cent market share at the airport. Boeing 737-800(W) LN-DYR heads a line-up on 4 October 2019.

On 31 May 2013, the demolition of Pier 1 began. This was Gatwick's second-oldest pier – the original 1962 south pier of what is now the South Terminal. It was replaced by a two-storey structure with five pier-served aircraft stands and an automated baggage-storage facility that cost nearly £200 million.

Today the airport is dominated by easyJet and TUI, with other airlines having a smaller presence.

The TUI Group is a global tourism leader, which emerged from a history rooted in German tour operators. Founded in 1968 as Touristik Union International, it evolved through mergers and acquisitions, including acquiring the Thomson Travel Group in 2000, which had already acquired Britannia Airways, and in 2007 merged with First Choice, a UK-based travel firm. This merger resulted in TUI Travel PLC, which later became TUI Group in 2014. The company has grown to encompass airlines, cruises, hotels, and travel agencies, solidifying its position as a leading provider of leisure travel experiences.

↑ Thomson Airways Boeing 737-400 G-CDZH on stand on 29 August 2006 prior to being fully integrated with TUI.

↓ Norwegian Air's Boeing 787 G-CKHL with an 'Amy Johnson' tail insignia on approach on 20 January 2018.

TUI's largest aircraft is the Boeing 787-8. G-TUIC is about to land on Runway 26 on 28 September 2019. (John Hughes)

TUI, as such, has been operating flights from Gatwick Airport since 21 June 2013, with the first service being a Dreamliner flight to Menorca. TUI uses both the North and South terminals at Gatwick, with North Terminal being for flights with the prefix TOM and South Terminal for flights operated by Titan Airways and Ascend with the prefix BY.

Another new airliner type appeared at Gatwick on 21 June 2013 when Thomson Airways operated the airport and airline's first Boeing 787 Dreamliner flight, a charter to Menorca. Thomson Airways was a subsidiary of the TUI Group, offering both scheduled and charter flights. Created in 2008 from the merger of Thomsonfly and First Choice Airways, Thomson Airways was rebranded as TUI Airways in 2017.

On 29 August 2014, a record 906 movements were recorded on Gatwick's main, equivalent to an aircraft taking off or landing every 63 seconds. It was considered to be the first time that more than 900 aircraft movements were handled by a commercial airport in one day using a single runway.

In the following twelve-month period, 40 million passengers passed through Gatwick for the first time. This was believed to be a global first for a commercial airport with a single-use runway. On 14 November, the 40 millionth passenger departed the airport on board a Norwegian Air Shuttle scheduled flight to San Juan.

On 1 February 2016, the redeveloped pier 5 of Gatwick's North Terminal was officially opened. This cost £80 million and formed part of the airport's £2 billion improvement programme under GIP ownership. It now features a vertical separation of departing and arriving passenger flows with an additional second level. This increases capacity by approximately 2,400 passengers per day or thirty flights. An additional seven large and twelve smaller aircraft (or a combination of both) can now be accommodated at the pier's redesigned aircraft stands, allowing greater operational flexibility.

In 2016 there was more expansion and improvements as on 14 June the new Pier 1 of Gatwick's South Terminal was officially opened, which cost £186 million. A new early bag store was included, which allowed up to 2,600 bags to be checked up to 18 hours before departure. Development of the new Pier 1 included new dual boarding facilities enabling boarding/disembarkation both via an airbridge or the aircraft steps, reducing queues at the passenger gates and speeding up the boarding/disembarkation process. Also a new lounge was opened for premium passengers on top of the new pier with panoramic views across the airport, which included four new taxiways and nine additional aircraft holding points.

By 25 January 2017, easyJet had consolidated all its Gatwick operations in the North Terminal, alongside Virgin Atlantic. British Airways moved into the South Terminal which improved the airport's operational efficiency and resilience. Different terminals used by easyJet and British Airways now reduced pressure on the North Terminal's check-in, security, boarding and ramp areas at peak times.

The number of passengers going through Gatwick continued to increase. In June 2017 it was recorded that 45 million passengers had passed through the airport for the first time on a rolling twelve-month basis. Regular scheduled flights were now offered to more than sixty long-haul destinations for the first time. This was believed to be a global first for a commercial airport with a single-use runway.

On 15 January 2018, the fastest transatlantic flight from New York to London by a subsonic passenger aircraft was carried out by a Gatwick-bound scheduled Norwegian Air Shuttle flight DY7014 from New York JFK with 284 passengers on board. This was operated by the airline's Boeing 787-9 Dreamliner G-CKHL featuring an image of British aviation pioneer Amy Johnson on its tail fin. This new record of 5 hours and 13 minutes was established as a result of tailwinds reaching a maximum speed of 202 miles per hour (325 km/h) over the Atlantic Ocean. The aircraft's maximum speed was therefore increased to 776 miles per hour (1,249km/h) and reduced the scheduled flying time by 53 minutes, enabling the previous record to be improved by 3 minutes!

↓ By 2001, Virgin Atlantic had revised their livery into this smart scheme as seen on Boeing 747-433 G-VROM on 30 March 2002.

In June 2018, a new five-year capital investment plan for the period until 2023 was announced which envisaged further funding £1.11 billion, taking the GIP's total investment since it bought Gatwick from BAA in 2009 to more than £3 billion. Of this additional expenditure, £266 million was allocated to the 2018–19 fiscal year. Several projects were to be covered, including an extension of the North Terminal's Pier 6 to the west. Also the relocation of the A380 stand to Pier 5, in conjunction with widening and reconfiguring the taxiway, enabling A380 operators to access Pier 5's new stand. Also included was a dedicated domestic arrival and baggage reclaim facility in the South Terminal to connect the new Boeing hangar with the airfield. By 2023, these planned improvements were designed to enable the airport to handle 53 million passengers.

Aiming for carbon efficiency, on 3 October 2018, Virgin Atlantic commenced a regular scheduled service to Gatwick from Orlando operated by a Boeing 747-400, which claimed to be the first commercial flight powered by a new type of sustainable aviation biofuel made by US-based LanzaTech from recycled waste carbon gases.

↓ ↗ More Gatwick history is recorded as Thomas Cook and My Travel, seen below on 29 August 2006, cease operations and disappear. This coincided with Virgin Atlantic being named 'Airline of the Year in 2001', as promoted on one of their Van Hool buses on a Gatwick–Heathrow shuttle run.

Thomas Cook Airlines, a British airline that operated for nearly two decades, ceased operations in September 2019 when its parent company, Thomas Cook Group, entered compulsory liquidation. The collapse resulted in the grounding of all Thomas Cook flights and the repatriation of hundreds of thousands of passengers.

→ Thomas Cook Airways Airbus A330 G-OJMC on stand on 26 May 2010. The company ceased operations on 23 September 2019 after Thomas Cook Group and its UK entities entered compulsory liquidation.

↓ On 2 November 2019, Air China launched its inaugural direct flight from Gatwick to Shanghai. Using Airbus A350-900XWB aircraft, the flights were operated four times a week. (Gatwick Airport Authority)

On 2 November 2019, Air China launched a four-times-a-week-service to Shanghai on Tuesdays, Wednesdays, Thursdays and Saturdays using an A350-900.

In 2020, the global COVID pandemic saw restrictions on international travel and flights were consolidated into one terminal, the first time that all flights operated out of a single terminal in nearly thirty years.

The following year, the South Terminal reopened for flights after eighteen months' closure due to the pandemic. Many airlines restarted flights as travel restrictions eased. Vueling opened a base at Gatwick and Norse Atlantic launched transatlantic flights to the USA with Boeing 787 Dreamliners. Wizz Air opened a base and JetBlue launched transatlantic flights to the USA.

↑ Another newcomer in 2022 was Norwegian airline Norse Atlantic Airways, which began flights on 12 August 2022. Their first route was between Gatwick and New York JFK using Boeing 787-9 Dreamliners such as G-CKWN, seen on 28 April 2025.

↓ On 18 May 2025, Uganda Airlines commenced direct flights between Entebbe and Gatwick, marking a major milestone in the national carrier's international expansion strategy. Airbus A330 5X-NIL is about to line up on runway 08 on that day. (Richard Vandervord)

In 2024, Gatwick was serving fifty-three long-haul routes including new routes to Singapore, Beijing and Guangzhou in China, Baku (Azerbaijan), Ashgabat (Turkmenistan), and Tashkent (Uzbekistan). In total the airport served nineteen destinations across Asia, including China, India, and the Middle East, thirteen in North and South America, sixteen in Central America and the Caribbean, and five across Africa. Also, twenty new retail outlets opened that year.

Gatwick continues to attract new airlines and new routes spanning many parts of the world. One example was on 18 May 2025, when Uganda Airlines launched a service to Entebbe that operates four times a week using Airbus A330-800neos. This aircraft has 258 seats across economy, premium economy and business cabins. Jonathan Pollard, chief commercial officer at Gatwick, said: 'This new route not only strengthens our position as a key gateway to Africa but also supports our vision of fostering economic growth and connectivity, particularly among new markets across Africa, Asia and the Middle East.'* These are the first direct flights between the UK and Uganda since British Airways ended its service in late 2015.

---

* www.mediacentre.gatwickairport.com/news/london-gatwick-grows-african-connectivity-with-only-uk-to-uganda-service-a35e1-40f32.html

→ South Terminal check-in desk area notification board.

EIGHT
GATWICK TODAY

In 2026, Gatwick celebrates its ninetieth anniversary, having come a long way from a small airfield alongside a race track in 1936. The airport's £2.2 billion expansion plan will include the redevelopment of the existing emergency runway and move it slightly further away from the sole main runway – meeting aviation safety rules to allow two runways to be used routinely at the same time.

Airline departure status board.

easyJet check-in area.

← The smaller airlines have not been forgotten. Welcoming staff at the Eastern Airlines and Aurigny Airlines check-in desks in April 2025. Since then UK regional airline Eastern Airways ceased all operations and entered administration in late October 2025, after twenty-eight years.

↓ Aurigny Airlines' colourful ATR-72 G-ORAI inbound to runway 26 on 26 September 2022. (Colin Cooke)

↑ One of the South Terminal departure zones check-in areas.

↗ The departures hall after customs and security. (Gatwick Airport Authority)

→ South Terminal main entrance and exit hall.

→ International arrivals hall.

↘ Entrance to the 'Peoplemover' travellator.

↓ South Terminal arrivals status board.

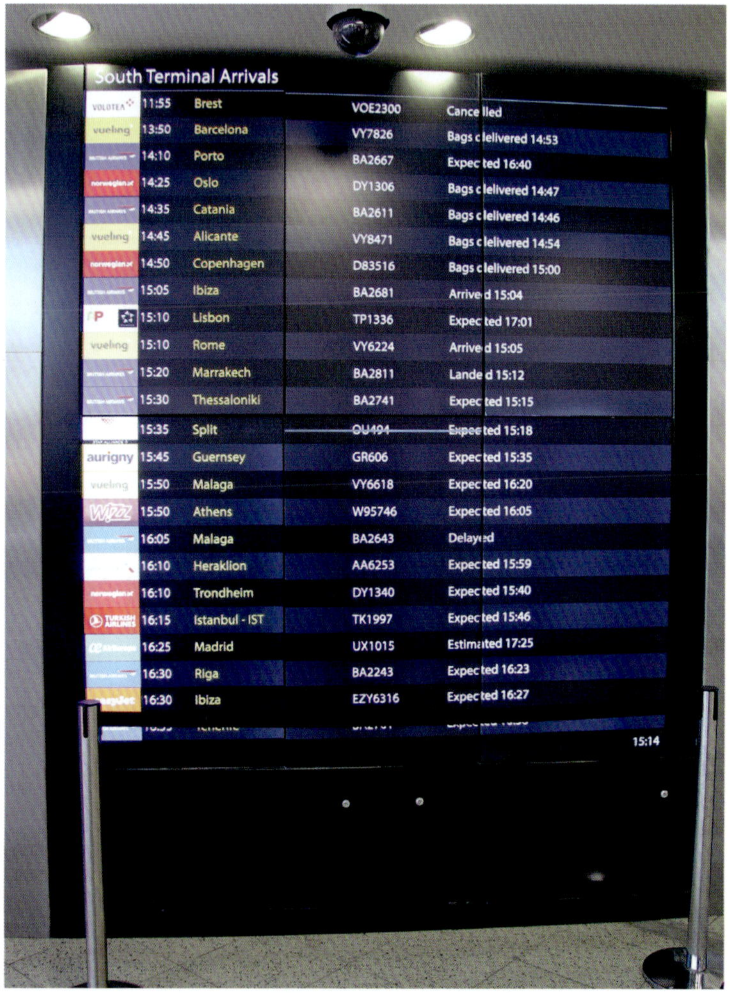

↑ The shuttle train running between the North and South Terminals.

↖ View from the shuttle train approaching the North Terminal.

← The North Terminal entrance. (Gatwick Airport Authority)

Aerial view of the control tower in April 2025.

↑ The North Terminal and connecting airbridge.

↑ Looking north towards Horley with the Hotel Sofitel and Premier Inn in the centre foreground.

The entire shuttle train track between the North and South Terminals can be seen here.

AERIAL PHOTOS BY IAN HASKELL

↑ Magnificent overview of the entire airport from the north-west.

↑ View looking towards the North Terminal with the cargo area hangars in the foreground.

↑ The airbridge connecting to the North Terminal.

The circular satellite pier extension that was opened in 1983.

↑ The North Terminal viewed from the north-west.

↑ We come full circle to where it all started in 1936: the original Beehive terminal building, still standing and now used as a conference centre.

ABOUT THE AUTHOR

(Simon Watson)

Adrian M. Balch has been an aviation enthusiast since the age of 8, having been encouraged and mentored by his late father. Born in Swindon in 1950, his first photographs were published in the late 1960s, followed by articles on military and civil aviation appearing in aviation enthusiast periodicals. He has had several books published, including this title's companion, *Heathrow in Photographs*, also published by The History Press. Adrian travelled extensively during his twenty-two-year RAF career in air traffic control, always with his cameras, and he has recorded aircraft worldwide for magazines and books. His first visit to Gatwick Airport was in 1968, not too late to see many interesting vintage airliners, and he has returned on numerous occasions to record the changes over the years.